SOFT CORALS AND SEA FANS

A comprehensive guide to the tropical shallow
water genera of the central-west Pacific,
the Indian Ocean and the Red Sea

by Katharina Fabricius and Philip Alderslade

Published by the Australian Institute of Marine Science,
PMB 3, Townsville MC, Queensland 4810, Australia

AUSTRALIAN INSTITUTE
OF MARINE SCIENCE

Australian
Biological
Resources
Study

CONTENTS

Copies available from:
Science Communication
Australian Institute of Marine Science
PMB 3, Townsville MC, Queensland 4810,
Australia.
www.aims.gov.au/softcoral.book
email: bookshop@aims.gov.au

The National Library of Australia Cataloguing-in-Publication data:

Fabricius, Katharina.
Soft corals and sea fans : a comprehensive guide to the tropical shallow water genera of the central-west Pacific, the Indian Ocean and the Red Sea.
Includes index.

ISBN 0 642 32210 4.

1. Alcyonaria - Indian Ocean . 2. Corals - Identification.
3. Alcyonaria - Red Sea. 4. Alcyonaria - Pacific Ocean.
I. Alderslade, Philip. II. Australian Institute of Marine
Science. III. Title.

593.6

Printed by New Litho, Surrey Hills, Melbourne, Australia
Finished by M&M Binders, Mount Waverley, Melbourne, Australia
Design, image processing and type setting by Katharina Fabricius

CONTENTS

LIST OF FAMILIES AND GENERA

ACKNOWLEDGEMENTS

Until now, no comprehensive and user-friendly reference material to soft corals and sea fans of the warm shallow waters of the broader Indo-Pacific region existed, yet there has been an urgent need for such a book. Most taxonomic descriptions are scattered throughout the scientific literature and fairly inaccessible for non-experts. Moreover, only recent scientific publications on soft corals and sea fans present underwater photographs of live specimens, whereas older publications depict preserved colonies that show little resemblance to the living forms in the field. We are very grateful to colleagues and friends who kept encouraging us to write this book, which we hope, will fill the void.

We are extremely grateful for funding and ongoing support from the Cooperative Research Centre for the Great Barrier Reef World Heritage Area (Reef CRC) in Townsville, the Museum and Art Gallery of the Northern Territory (MAGNT) in Darwin, and especially from the Australian Institute of Marine Science (AIMS) in Townsville. All three Institutions supported this project by giving us time to work on this book. Many thanks to Terry Done for his generosity and ongoing scientific support and friendship. A large proportion of the underwater photographs presented here were compiled as a reference collection during octocoral surveys of the Great Barrier Reef, funded by the Reef CRC and AIMS. AIMS provided all ship time and thus access to the Great Barrier Reef, and both MAGNT and AIMS provided the research facilities. We would not have been able to produce this book without a generous grant by the Australian Biological Resources Study (ABRS) of the Department of Environment, Australia, and we would like to particularly acknowledge the ongoing support of taxonomic and biodiversity related research in Australia facilitated by the ABRS. A research grant by the SWIRE Marine Laboratory in Hong Kong enabled us to expand the coverage of this book into the South China Sea, and to collect many additional photographs from that region.

We would like to greatly thank Gary Williams, Frederick Bayer, Russell Kelley, and Leen van Ofwegen who reviewed a late draft of the manuscript. We are immensely grateful for their expert advice and their many valuable suggestions for improvement. We would like to particularly acknowledge the contributions of Leen van Ofwegen who acted as a sounding board for numerous aspects of this book. Our taxonomic understanding of the family Melithaeidae and several genera of the family Nephtheidae benefited greatly from Leen's ongoing research, and he generously helped us with the descriptions of these two families, and the sclerite drawings of the genera *Nephthea, Stereonephthya,* and *Litophyton.* All other sclerite drawings were done by Phil Alderslade.

This book endeavours to cover octocoral genera that can be found over a vast area of the world, and within each genus we have attempted to show a variety of different growth forms. In order to do this we sought help from valued friends, and we thank and acknowledge their generosity in contributing numerous valuable photographs. They are, in alphabetic order: Carolina Bastidas, Fred Bavendam, Günther Bludszuweit, Robert Bolland, Pat and Lori Colin of the Coral Reef Research Foundation, Lyndon Devantier, Frances Dipper, Graham Edgar, Harry Erhardt, Doug Fenner, Karen Gowlett-Holmes, Manfred Grasshoff, Bert Hoeksema, John Hooper, the Marine Bioproducts Research Group of the Australian Institute of Marine Science, Dick Mariscal, Kirsten Michalek-Wagner, Alf J. Nilsen (Bioquatic Photo, bioquatic@aquariumworld.com; www.aquariumworld.com/bioquaticshop), Leen van Ofwegen, Gustav Paulay, Gordon LaPraik (Senior), Kevin Reed, Götz Reinicke, Matt Richmond, Michael Schleyer, Julian Sprung, Roger Steene, Ann Storrie, Tina Tentori, Emre Turak, the UNEP World Conservation Monitoring Center, Gary Williams, and Mark Wunsch. The authors are acknowledged individually under each photograph, and the copyright for these pictures remains with them.

We thank the crews of the AIMS Research Vessels *Harry Messel, Cape Ferguson* and *The Lady Basten* for their tremendous support in carrying out the surveys and research for many years. Many volunteers, too numerous to list here, are also greatly thanked for their help with the field work over the years. We would like to particularly acknowledge the contributions of Gordon LaPraik (Senior), who has observed and recorded octocorals and their molluscan symbionts in Keppel Bay (Central Queensland) since 1955. His offer to use his collection of underwater photographs was one of the incentives to proceed with producing the book.

We thank Mary Stafford-Smith and Charlie Veron for their encouragement, and their help with technical aspects of publication and layout. Similarly, Michael Browne, Liz Tynan, Steve Clarke, Natalie Daly, Nadene Jones, and Klaus Uthicke helped in various ways with layout and technical aspects of the production. Russell Kelley, Kirsten Michalek-Wagner, Simon Moore and Elaine Robson provided helpful scientific advice on parts of the introduction. They all deserve our sincere thanks.

Lastly, we greatly thank Ann Alderslade for her great help in proof-reading the book. And above all, Glenn De'ath has been a wonderful source of creative ideas and encouragement. The book wouldn't be what it is without him.

Sinai, northern Red Sea. *Photo: KF*

FOREWORD

The Indo-West Pacific is home to the richest, most diverse octocoral fauna of world seas. By current reckoning, some 23 families comprising 90 genera, most with several or many species, inhabit waters accessible to skin- and SCUBA divers. These organisms are vital components of shallow water and reef environments throughout the area, some even contributing substantially to reef structure. Some are significant sources of natural products of real or potential importance in biomedical research, pharmacology, and cancer therapy. They are of intense interest to biochemists and ecologists as well as to systematists concerned with their identification, classification and phylogeny. The phenomenal increase in research activity devoted to Octocorallia in recent years testifies to the timeliness of this book, and to the importance it will have in the scientific community unfamiliar with taxonomic investigation and with research methods required even for preliminary identification.

The proportion of taxonomic papers to other aspects of octocoral biology clearly reflects the low esteem in which the scientific community regards taxonomy at the present time. As reported by the Zoological Record for 1950, 8 of 16 papers on octocorals by 11 authors were taxonomic. In 1998-99 the ratio was only 7 of 81 papers by 69 authors. Researchers investigating non-taxonomic aspects of octocoral biology are under constant pressure to produce results, so most are unwilling or unable to delay publication pending verification of the identity of their animals by specialists, who may require weeks or months to provide reliable determinations. The validity of results of research performed upon species of questionable identity could eventually be brought into question or repudiated. Only 5 authors dedicated to the taxonomy of octocorals have been active in the past decade, and one of those has now retired from active research. An additional number of academics have contributed to the solution of limited taxonomic problems, but they by no means satisfy the widespread need for taxonomic expertise.

This volume, written in informal, non-technical prose, will go a long way toward ameliorating the difficulty of identifying octocoral genera used in research on a wide variety of aspects of their biology by non-taxonomists. It provides a comprehensive introduction to the anatomy, biology and classification of Octocorallia, as well as a guide to collection, preservation, and preparation of specimens for study. The authors have assembled photographs of living examples of essentially all genera of Octocorallia known from the shallow waters of the Indo-West Pacific and Red Sea. Together with information about numbers of species, habitats and distribution, illustrations of essential taxonomic features characteristic of diverse genera, this book provides non-taxonomists with a comprehensive guide to generic level identification of all shallow water octocorals now known to inhabit the geographic regions involved. It will serve as a point of departure towards the solution of taxonomic problems that have plagued biologists and systematists for the better half of a century. So many imperfectly known species are represented that a species-level taxonomic treatment was unfeasible, and will probably remain unfeasible for the foreseeable future. But this book, a first in its breadth of coverage, will remain a cornerstone of information about the genera and families of the Indo-West Pacific Octocorallia for decades to come.

Dr. Frederick Bayer

Zoologist Emeritus
Smithsonian Institution, Washington

Top: Madang, Papua New Guinea. *Photo: Roger Steene;* Bᴏᴛᴛᴏᴍ: Off-shore reef, central Great Barrier Reef. *Photo: KF*

INTRODUCTION

About this Book

This book is a guide to families and genera of soft corals and sea fans from the shallow, tropical and subtropical regions of the Indian and Central-West Pacific oceans and the Red Sea. Soft corals and sea fans are common names for species of animals grouped under the scientific name Alcyonacea. Together with blue coral and sea pens, they make up a larger animal group called Octocorallia (BOX 1). Their distinguishing characteristic is that their polyps always bear eight tentacles (hence octo-coral), which are fringed by one or more rows of pinnules along both edges. The popular term "soft coral" points to the fact that most soft corals, in contrast to the related hard corals, have no massive solid skeleton.

Octocorals occur in virtually all marine realms. They are found in every ocean, from the tropics to the poles. Species live in habitats ranging from the intertidal through brackish muddy estuaries to oceanic blue water and abyssal depths. The shallow waters of the Indonesian-Philippine-New Guinea archipelagos have recently been shown to host the greatest number of octocoral species. From this centre of octocoral biodiversity, species numbers decline towards higher latitudes, and towards the eastern and western rims of the Indo-Pacific. Cool or deep regions are also inhabited by fewer species.

Around 90 genera of Alcyonacea, belonging to 23 families, have been described from the tropical Indo-Pacific at diving depth (see List of Families and Genera, page iv). This compares with about 89 genera and 18 families of Indo-Pacific hermatypic hard corals (Veron 2000). In this book we have produced a tool which, for the first time, should give non-experts a comprehensive overview of these 90 genera, and the means to correctly identify them under water and in the laboratory. We have used the taxonomic system as it is presently accepted. The sequence of taxa largely follows Bayer (1981), however, we have incorporated numerous updates that have been published in the meantime. The evolutionary relationships between taxa remain largely unknown, so we have arranged the genera within families by the similarity of their features. Future taxonomic revisions will undoubtedly necessitate changes in these arrangements.

BOX 1: Clarification of the Nomenclature:

The term **soft coral** has different meanings in different parts of the world. Most commonly this term is used to refer only to Octocorallia which have no massive skeleton or internal axis, but sometimes it also includes the sea fans, and occasionally it is used to refer to all Octocorallia. Similarly, the term **sea fan** is generally used as an alternative for **gorgonian**, which are Octocorallia other than sea pens, which arise from the substrate with the support of an internal axis. It is not unusual, however, for people to use it exclusively for gorgonians with a fan-shaped morphology, contrasting them with unbranched species, which they generally call "sea whips" or "sea rods". It is important to point out that there is no distinct dividing line between soft corals and gorgonians (see also FIG. 8), although the terms are useful to retain. Twenty years ago scientists classified them into a number of different groups. Research into the structure and growth of octocorals has bridged the gaps between these groups and they are now all included under the one scientific category – the **Order Alcyonacea**. All species in this group can correctly be referred to as alcyonaceans, but in this book we refer to them, as we would in conversation, as **octocorals**. Although this term can include blue coral and the sea pens, we use it to include only soft corals and sea fans, unless stated otherwise.

We have minimised the use of technical terms to make the guide as user-friendly as possible without compromising accuracy. The book should therefore be of interest to at least four types of readers:

- marine scientists and marine biology students interested in the biology, ecology and taxonomy (identification) of octocorals;

- people involved in surveying and monitoring coral reefs and other shallow-water sea floor habitats;

- scuba diving and snorkelling underwater explorers of tropical waters; and

- marine aquarists and other naturalists who would like to name and know more about the animals they care for.

Although we have reduced jargon as much as possible, some technical terms were unavoidable, however we attempt to explain each as we go along in the introduction and the text throughout. The **pictorial guide to the families and genera**, which precedes the detailed descriptions, should also help with identification by quickly narrowing down the range of groups to which a specimen may belong.

In describing the individual genera, we present three sources of information:

- **underwater photographs** of colonies and a range of growth forms. Close-up images highlight certain characteristic or diagnostic features. Preserved specimens are used to illustrate some of the rarer octocorals, or features not clearly visible in live colonies;

- the small skeletal elements that occur in the tissues of colonies (formally referred to as spicules, but more correctly called sclerites) are represented by drawings. The **figures of sclerite drawings** have been composed from several different species in an attempt to show the range of the different shapes that may be encountered in a genus. Most soft corals and many gorgonians can be identified to genus level in the field by features that can be seen with the unaided eye, however, other genera require a microscopic investigation of sclerites and other structures in order to establish their identity;

- **detailed text** describes the **Colony shapes**, **Polyps**, and **Sclerites**, as well as **Colouration** and other notable characteristics of each genus. It also summarises the present knowledge of their ecology and natural history, habitat preferences, abundance, distribution patterns, and zoogeography. Information about **Habitat and abundance** is very scarce in most parts of the world, therefore our descriptions are mostly based on own observations from the Great Barrier Reef (GBR) and it has to be kept in mind that it may differ in other regions. Similarly, the information on **Zoogeographic distribution** ranges is based on a limited number of records, as surveys and species inventories of octocorals exist only from a few locations. Correspondingly, most cited ranges are likely to be incomplete, and will need to be revised once more records are available. The description of each genus finishes with a list of **Similar taxa** from the regions covered by this book, to additionally aid comparison and identification.

An Overview of the Classification System

Coelenterata

The phylum Coelenterata (see Box 2) contains tentacle-bearing invertebrates (animals without backbones) with stinging structures, more or less radial symmetry, and only one major internal cavity (the digestive cavity). This cavity has only one opening: the mouth. A unique feature of the phylum is the possession of stinging structures, called nematocysts. Two main growth stages are found in Coelenterates: the polyp, and the medusa.

- The **polyp** is cylindrical in shape, with one end attached or resting on the substratum. The mouth is at the free end, and usually surrounded by tentacles (FIG.1A). Corals, both soft and hard, are formed either by a single ("solitary") polyp, or by a colonial arrangement of many polyps, which cooperate as "building blocks" to make modular colonies.

- The **medusa**, commonly known as a jellyfish, consists of an umbrella-, bowl- or bell-shaped body that is sometimes fringed by tentacles (FIG. 1B). Most medusae are free-swimming, however a few forms live attached to the substratum.

In both forms, the body wall consists of two types of tissue: the outer layer (**epidermis**) bears the sensory and stinging cells, whereas the inner layer (**gastrodermis**) is responsible for digestion and reproduction. A third layer, called the **mesogloea**, particularly well developed in jellyfish, connects these two layers. Depending upon the particular animal group, mesogloea consists of material ranging from cell-free cement to a thick gelatinous substance that can contain fibres and cells and skeletal elements. In octocorals it is called the coenenchyme. In its more advanced and complex form, mesogloea has all the characters of a true connective tissue.

Coelenterates are the most primitive animals to have regular muscle and **nervous systems**. They possess circular, radial, and longitudinal muscular strands and layers, allowing for a limited range of motions including pumping, contraction and bending. Their nervous system consists of one or two networks with diffuse neurons and sensory cells. The sensory cells act as unspecialised general receptors for stimuli such as touch, chemicals, or sudden change in temperature.

The phylum Coelenterata is probably around 500 million years old. Extant coelenterates are divided into four classes, and a large number of orders and families. The list on the following page provides an overview of the systematic structure of the phylum. Other texts may offer slightly different versions. The group treated in this book is underlined. Where they exist, common names are given together with the scientific names.

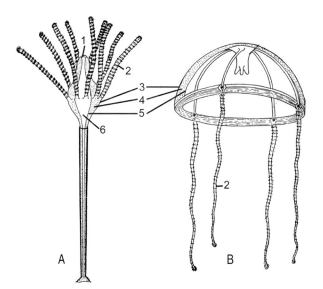

FIG. 1: The main growth stages found in Coelenterates: **A.** polyp (here: a stylised hydroid polyp), and **B**: medusa (here of a hydroid). 1 = mouth, 2 = tentacles, 3 = epidermis, 4 = mesogloea, 5 = gastrodermis, 6 = gastro-vascular cavity. Modified from Hyman, 1940.

· **Class Hydrozoa:** This class includes the hydroids, the milleporine and stylasterine corals, together with some medusoid life stages. Both polyp-bearing and medusoid forms are represented among the approximately 7 living orders of this class. Most are marine, but there are also some freshwater forms. Around 2700 species have been described.

· **Class Scyphozoa:** These are the jellyfish. All are medusoid forms, and most are free-swimming, while others are attached by a stalk at least during part of their development. Most species in this class are marine, a few inhabit freshwater. Five orders and around 200 species have been described.

· **Class Cubozoa:** This is a small group of marine jellyfish with a cubic bell shape, four flattened sides, a domed top, a square transverse section and individual tentacles or tentacle bundles at the four corners. A famous example is the deadly box-jellyfish, *Chironex fleckeri*.

· **Class Anthozoa:** This class contains soft and hard corals, sea pens, anemones, zoanthids, corallimorpharians, tube anemones and black corals. It has only polypoid forms (no medusoid stage), and polyps are solitary or colonial. The polyps have a gastric cavity that is subdivided by longitudinal partitions (septa or mesenteries). There are three subclasses, all of which are marine throughout their lives, and together contain 9 orders and around 6000 extant species.

> · Subclass Octocorallia (or Alcyonaria)
> > · Order Helioporacea (blue coral)
> > · Order Alcyonacea (soft corals and sea fans)
> > · Order Pennatulacea (sea pens)
>
> · Subclass Hexacorallia (or Zoantharia)
> > · Order Actiniaria (sea anemones)
> > · Order Scleractinia (hard corals)
> > · Order Zoanthiniaria (zoanthids)
> > · Order Corallimorpharia (mushroom anemones)
>
> · Subclass Ceriantipatharia
> > · Order Antipatharia (black and wire corals)
> > · Order Ceriantharia (tube anemones)

Some fundamental characteristics distinguish the three anthozoan subclasses. The most pronounced feature is that in the subclass Octocorallia, each of the polyps bears eight hollow tentacles, which are fringed on both sides by one or several rows of pinnules (FIG. 2 AND 5), whereas the polyps of the subclass Hexacorallia bear six, or multiples of six tentacles, without pinnules. Ceriantipatharia includes the black corals and wire corals, and anemone-like, cylindrical, sand-dwelling anthozoans. Antipatharians have a slender growth form, commonly branching (not unlike the gorgonians), a central skeletal axis covered with thorns, and short, cylindrical, non-retractile polyps bearing six tentacles.

FIG. 2: The polyp of an octocoral is characterised by bearing eight tentacles, fringed by rows of pinnules. Here: a colony of *Tubipora*, view: ca 4 cm. *Photo: KF*

BOX 2: Classification of organisms:

In order to group the vast number of species by their similarities and relationships, taxonomists (those who study the identification of life forms) established a hierarchical system of categories. Seven basic categories are used in sequence to describe increasing levels of similarity in attributes which are of diagnostic value. Additional intermediate categories, such as **tribe** exist, and others are inserted by adding the pre-fix 'sub-' (e.g., subclass), '**infra-**' (eg. infraspecies), or '**super-**' (eg. superfamily):

- **Kingdom:** This is a category of primary division that distinguishes animals from viruses, bacteria, protozoans, plants, and fungi.

- **Phylum** (plural: phyla): This is the highest basic category below Kingdom. It encompasses groups with a long joint evolutionary history and profound similarities in basic structures and developmental characteristics. An example is the **Phylum Echinodermata**, which includes all starfish, sea cucumbers, sea urchins, feather stars and other related forms. Presently, there are 38 phyla described, 34 of which contain marine forms. Each phylum is split into classes.

- **Class:** Classes are major categories of primary distinction. An example is the **Class Echinoidea,** which only includes the sea urchins. Each class is divided into orders.

- **Order:** This is the next level down, and is a category that encompasses a group of closely related families. An example is the **Order Clypeasteroida,** which comprises all the flat sea urchins commonly called sand dollars.

- **Family:** in the animal kingdom, the name of a family always ends with "…**idae**". An example is the **Family Mellitidae,** which contains just the arrowhead sand dollars and the key hole sand dollars. Families are divided into genera.

- **Genus** (plural: genera): This category represents a group of closely related species. An example is the genus **Encope**, which incorporates only the arrow sand dollars.

- **Species:** The most specific rank in the Linnean hierarchy – sometimes subspecies are separated out next. A species, by traditional definition, is a group of organisms which can freely interbreed and produce fertile offspring, but is reproductively isolated from other species (more recent research indicates that this is not always true, as some coelenterates produce fertile hybrid offspring). An example of a species is **Encope michelini**, a dark green arrow head sand dollar that is common in south Florida and the Caribbean.

A species name, by convention written in *italic* font, always consists of two parts: the name of the genus to which it belongs is given first, followed by a unique identifier for the species. For example, *Sinularia flexibilis* is one of more than 120 soft coral species within the genus *Sinularia*. The species name is often derived from Latin or Greek, and commonly points to a characteristic feature of that species, names the place of first collection, or honours a person. In *Sinularia flexibilis*, the name reflects the distinctive, flexible and long branches characteristic of this species.

Octocorallia

Among the Octocorallia, the **Helioporacea**, **Pennatulacea**, and **Alcyonacea**, are clearly delineated as three distinctly separate orders. The age of the Octocorallia, and time of evolutionary separation between the three orders is unknown, because pennatulacean and alcyonacean sclerites are small and quickly wash away after a colony dies, so fossil records are very sparse. Thus the early evolution of the octocorals may never be reliably resolved.

Helioporacea

The only extant shallow-water species of Helioporacea, the blue coral *Heliopora coerulea* (FIG. 3), is a living relict, with closely related fossil species known from more than 100 million years ago. It is the only octocoral that forms a massive aragonite skeleton, like hard corals, or fire corals (*Millepora*). The blue colouration of its skeleton is most clearly visible on broken branches, because a thin layer of brownish live tissue envelops the outer surface of the skeleton. The delicate polyps are housed in cylindrical pores in the skeleton, and are connected through smaller pores called solenial tubes. Additional solenia traverse the thin layer of surface tissue. Both types of skeletal pores are closed off at the bottom end by a transverse plate. As the colony grows, a new traverse plate is formed, moving the living tissue into the newly created "upper floor" and effectively closing off the formerly inhabited lower floor. The species contains zooxanthellae and is therefore dependent on light. Its distribution is restricted to shallow-water areas in which the water temperature remains above 22°C all year round. The order additionally contains the family Lithotelestidae, with four known species from deeper waters of the Atlantic Ocean and Madagascar, which are small and stoloniferous (Bayer 1992).

FIG. 3: The blue coral *Heliopora coerulea*. **A:** Cross-section and surface of a dried skeleton, view ca 4 cm. **B:** A colony from Palau. **C:** Large colonies are found in shallow near-shore waters of the Great Barrier Reef. *All Photos: KF*

Pennatulacea

The sea pens or Pennatulacea are a diverse but poorly known group of octocorals containing 16 families (Williams 1995). The majority of pennatulaceans live in the deep sea, but many species can also be found in shallow waters of the Indo-Pacific (Fig. 4). However, because they live in soft-bottom habitats and many are often completely withdrawn into the substrate during daylight, they are only infrequently encountered by divers and snorkellers. Taxonomic reviews by Williams (1992a, 1995, 1999) are recommended for their identification. Pennatulaceans are characterised by their large, central, primary or **axial polyp**, called the **oozooid**, which is usually supported by an internal calcareous axis. Half of the oozooid forms the colony "foot" or **penduncle** that digs into sand or mud, anchoring the colony in soft substrata. This part is permanently hidden from view. The second half of the polyp, the **rachis**, reaches into the water column when expanded, and bears numerous other polyps, called **autozooids** and **siphonozooids**. In some species, the emergent part looks just like a feather, hence the name sea pen, but the shape and colouration of pennatulaceans varies considerably. Many sea pens have sclerites, which can vary from minute platelets to long needles. Some species have large, acutely pointed sclerites arranged on the outer edges of the colony for protection.

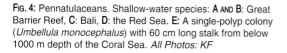
Fig. 4: Pennatulaceans. Shallow-water species: **A** AND **B**: Great Barrier Reef, **C**: Bali, **D**: the Red Sea. **E**: A single-polyp colony (*Umbellula monocephalus*) with 60 cm long stalk from below 1000 m depth of the Coral Sea. *All Photos: KF*

Alcyonacea

Of the 29 families in the order Alcyonacea, 23 are found in the warm, shallow waters of the Red Sea, the Indian and Central-West Pacific oceans (see page iv). Just as living animals undergo evolution, so does the system of classifying them as additional knowledge is discovered. At present, the system is in its least complex stage, with all soft corals and gorgonians being placed in the single order. The past has seen the families separated into groups under such orders as: Protoalcyonaria, Stolonifera, Telestacea, Gastraxonia, Xeniacea, Alcyonacea and Gorgonacea. The establishment of Gastraxonia and Xeniacea did not receive much support, but the remaining five orders formed a relatively practical framework that was used for 50 years. However, over the years, as previously described species were examined in more detail, and as new species were discovered, it became increasingly clear that intermediate forms between these ordinal groups prevented the definition of any clear boundaries. The species we recognise today form a complete series from simple soft corals to complex gorgonians (see Fig. 8, page 11). In 1981, Bayer proposed a far more realistic system: a single order, Alcyonacea. This system has been adopted by most modern taxonomists, and is the one followed in this book. It should be pointed out that the order Alcyonacea does actually have discontinuities within it. There are two groups of complex gorgonians, the Holaxonia and the Calcaxonia, that differ in axis construction where no intermediate forms have yet been found. They are included as suborders, bacause they are not considered sufficiently different to be classified as separate orders.

BOX 3: The root Alcyon...:

The root "Alcyon…" occurs at five different levels in the octocoral literature, and often creates confusion:

- **Alcyonaria** is an outdated, but still occasionally used, name for the <u>subclass</u> Octocorallia. The adjective is **alcyonarian**.
- **Alcyonacea** is the main <u>order</u> of octocorals covered in this book; the adjective is **alcyonacean**.
- **Alcyoniina** was proposed as a <u>suborder</u> of Alcyonacea to separate out the more massive species. But intermediate forms blur its boundaries, and the name has rarely been used. We retain it here as a group name.
- **Alcyoniidae** is one of the three largest soft coral <u>families</u>, which includes, for example, the genera *Sinularia*, *Sarcophyton* and *Lobophytum*; the adjective is **alcyoniid**.
- *Alcyonium* is one of the <u>genera</u> belonging to the family Alcyoniidae. It is found in the Atlantic and eastern Pacific Oceans, whereas no valid species of this genus lives in tropical waters.

Biology of Octocorals

Histology and Anatomy

The Polyps

Most octocorals have only one type of polyp, the autozooid, which in the majority of cases is responsible for food capture and reproduction (Fig. 5A). Species with only one type of polyp are termed monomorphic. A few species, mostly larger forms, are called dimorphic because they have a second kind of polyp called a siphono-zooid, which is smaller, and has no, or rudimentary, tentacles (Fig. 5B and 6). Their primary function is supposedly to irrigate colonies with seawater, however dissolved and small suspended food particles may also be transported into the colonies along with the water.

Fig. 5: The structure of octocoral polyps. **A:** Two autozooids of a *Clavularia* sp. (here: a mature and a newly budded, yet fully developed smaller polyp). View: ca 4 cm. **B:** A *Lobophytum* sp., showing the numerous openings of the small siphonozooids as darker dots around the bases of the larger and fully developed autozooids. View ca 2.5 cm. *Photos: A: Roger Steene, B: KF*

An autozooid is essentially a cylindrical or tubular structure with a mouth and tentacles at one end (FIG. 5 AND 6). That portion of the polyp that protrudes into the water encloses a space called the **gastric cavity** or **gastro-vascular cavity**, and has a wall that is 3 layers thick. The upper, free end of the polyp has a centrally situated mouth opening, which is surrounded by eight **tentacles**. These tentacles have finger-like extensions along each side, called **pinnules**, which give them a feathery appearance and greatly enhance both the inner and outer surface areas of the polyp. Pinnate tentacles are mobile and contractile, densely covered with sensory cells and stinging capsules, and often filled with symbiotic algae. They are thus are equipped to both sense and grab an impacting particle. The number of pinnules per row, and the number of pinnule rows on each side of the tentacle, varies between species. In a few species the pinnules are so small as to be nearly invisible, and in some newly discovered forms, not yet described, the pinnules are fused together.

Extending down from the mouth into the gastric cavity is a short tube called the pharynx (FIG. 6 AND 7). This tubular structure is oval in cross-section, and one of the narrow sides is formed into a groove, termed the **siphonoglyph** or sulcus. It is lined with ciliated cells that serve to drive

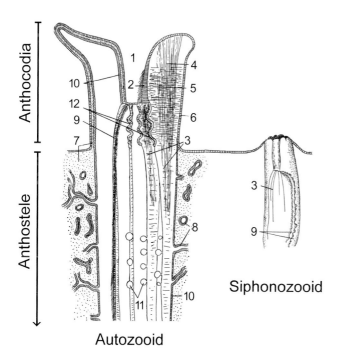

Anthocodia

Anthostele

Autozooid

Siphonozooid

Fig. 6: Schematised representation of an autozooid and a siphonozooid. 1 = pharynx, 2 = siphonoglyph, 3 = septa, 4 = retractor septal muscle, 5 = transverse fibres of mesentery, 6 = epidermis, 7 = mesogloea, 8 = solenia, 9 = asulcal mesentery, 10 = gastrodermis, 11 = gonads, 12 = mesenterial filaments. Modified from Hyman, 1940.

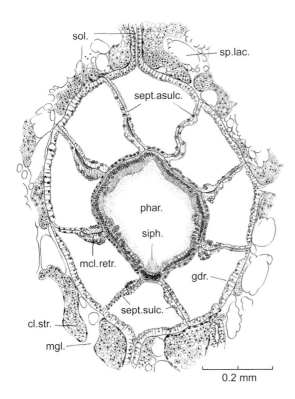

FIG. 7: Cross-section through an autozooid polyp. Phar. = pharynx, siph. = siphonoglyph, sol. = solenium, sept. asulc. = asulcal septum, sept. sulc. = septum, gdr = gastrodermis, mgl. = mesogloea, sp. lac = lacuna remaining after solution of spicule, mcl.retr. = retractor muscle. cl.str. = cell strands. After Bayer, 1974.

0.2 mm

water into the gastric cavity of the polyp, and from there to the rest of the colony via a canal system. It is exceptionally well developed in siphonozooids. Extending into the gastric space from the walls of the polyp are eight thin, radially arranged mesenteries (FIG. 6 AND 7). The upper, inner edges of the mesenteries are connected to the walls of the tubular pharynx, while below this the edges are free. The parts of the gastric space around the outside of the pharynx, which are partitioned by the mesenteries, extend up into the tentacles and pinnules, which are hollow.

BOX 4: Polyp Anatomy

The outer tissue layer of the polyp, where it extends into the water, is called the **epidermis** (FIG. 6). It contains numerous mucus-producing cells, sensory cells, microvilli, and various numbers of stinging capsules (nematocysts, see Box 5). In expanded octocorals, the outer surface is increased many-fold by the eight tentacles and numerous pinnules that surround each polyp. The tissue layer that covers the internal mesenteries, and lines the pharynx, gastric cavity and tentacles, is called the **gastrodermis**. Between the epidermis and the gastrodermis is the third layer called the **coenenchyme**. It consists of a gelatinous material containing some fibres, amoeboid cells, and scleroblast cells along with the calcareous sclerites that they have produced. The internal surface area (gastrodermis) is greatly enlarged by eight protruding tissue plates (the **mesenteries**), and by a network of numerous narrow canals (called **solenia**), which traverse the fleshy coenenchyme between the polyps (FIG. 7). Mesenteries have an important function in this hydraulic system, as without them an inflated polyp would bulge like a balloon instead of remaining cylindrical. Although the coenenchyme is very thin in the walls of the polyps, it represents the main bulk of the colonies. In thick colonies, a few or many of the polyps have the gastric cavity continued deep into the colonial mass as a long tube called the **gastrodermal canal**. This canal is similarly connected to other parts of the colony by solenia. All of these tubes and canals are lined with gastrodermis. In some octocorals where the colony is only thinly layed upon an axis or the substrate, the gastric cavity is very short, and is connected to other parts of the colony by solenia. Water is propelled through this network of canals by the synchronised beat of cilia, supplying the inner tissues with dissolved gases, and transporting nutrients between polyps.

The free edge of each mesentery is thickened into a long cord called a **mesenterial filament**. The two filaments on the side opposite from the **siphonoglyph** are the longest and thickest, often bilobed, heavily flagellated, and usually continue to the very base of the polyp. The filaments on the other six mesenteries contain digestive gland cells, and in most genera also produce the gonads. In one family (Xeniidae) these six filaments are reduced or rudimentary, as is also the case in siphonozooids.

An octocoral polyp consists of two main parts (FIG. 6). The whole portion that is able to contract, extend, and wave about above the colony surface is called the **anthocodia** (plural: anthocodiae). It consists of the cylindrical body together with the terminal mouth and tentacles. This part of the polyp may be able to completely retract from view. The lower part of the polyp is called the **anthostele**, but the term is rarely used, especially in the correct way. The anthostele is the whole gastrodermal canal that extends down from the anthocodia, penetrating the common colonial mass, or stolons, and in most cases it is invisible without dissecting the colony. It is essentially the layer of gastrodermis that lines this single canal, together with any mesenterial structures within the canal. In many species, by a process of invagination in the "neck" region of the polyp body below the level of the pharynx, the anthocodia can totally **retract** (e.g., FIG. 9G, H, J, L, M) within the upper part of the anthostele. In other species, polyps **contract,** by simpe deflation without invagination (e.g., FIG. 9E, F). In a third group of species, polyps can neither retract nor contract. In some species, the upper part of the anthostele extends above the colony surface within a sclerite-reinforced mound called the **calyx**. An anthocodia that extends above a calyx is always able to retract within it. In some stoloniferous species of soft coral, the anthostele is mainly contained within a very long calyx, ending as just a small depression in the stolon below.

The Colonies: Growth Forms, Sclerites and Axes

While the shape of polyps is remarkably constant in all octocorals, the shape and size of colonies (which may have a few or many thousands of polyps) can vary widely from one species to another. Moreover, colony shape can also vary considerably within certain species, sometimes in response to environmental conditions such as light availability and wave exposure. Colony **growth forms** are divided into broad categories: membranous, encrusting, massive, lobate, digitate, arborescent, fan-shaped, bushy, and unbranched or whip-like growth forms are perhaps the most common categories, however large numbers of intermediate forms exist. Definitions of the various growth forms, and many other commonly used technical terms, can be found in the "Illustrated trilingual glossary of morphological and anatomical terms applied to Octocorallia" by Bayer *et al.* (1983).

The simplest octocoral is a single polyp. *Taiaroa tauhou* from New Zealand is the only confirmed single-polyp species. *Umbellula monocephalus* (FIG. 3E) and *Anthomastus robustus* have a single autozooid, but also siphonozooids. Other taxa have been described as single-polyp species, however in reality the specimens may have been founder polyps, collected before they had begun forming a colony. In the families Cornulariidae and

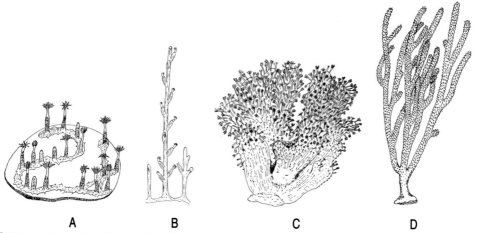

A **B** **C** **D**

FIG. 8: Various octocoral growth forms: **A:** multiple polyps connected by stolons (e.g., genus *Clavularia*); **B:** tall, axial-polyps that produce secondary polyps from their sides (e.g., *Carijoa*); **C:** in soft corals, anthosteles are embedded in the common, more or less fleshy, coenenchymal mass of the colony (e. g., *Efflatounaria*); **D:** gorgonian growth form (e.g., Plexauridae). After Hyman, 1955.

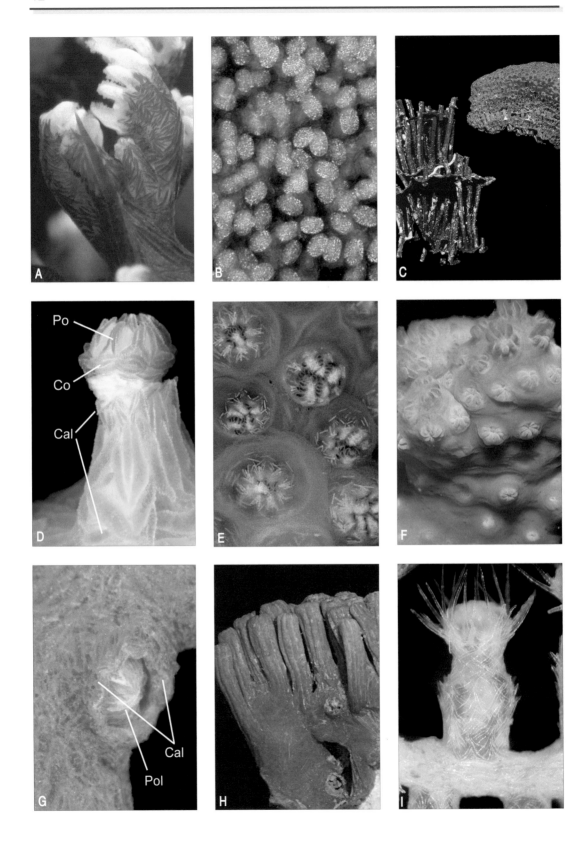

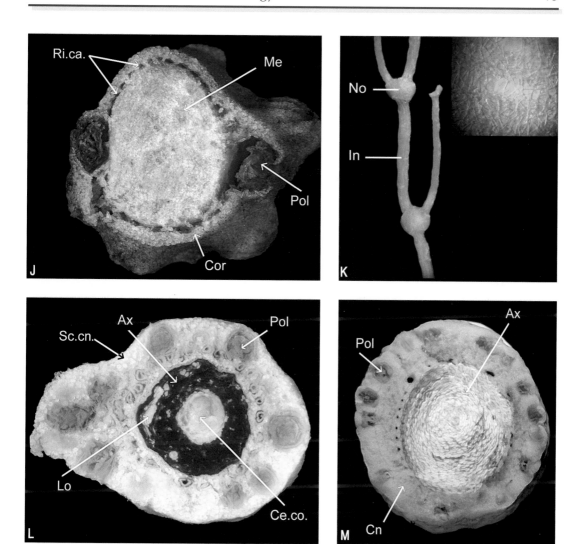

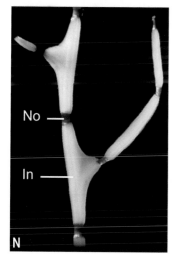

Fɪɢ. 9: Skeletal structures in octocorals. **A: Supporting bundle**, and loose sclerites in the translucent tissue of a branch of *Dendronephthya* (7 mm). **B:** Loose sclerites in the translucent tissue of a xeniid (here: *Sansibia*, 0.2 mm). **C:** The skeleton of the organ-pipe coral *Tubipora* sp. formed from fused sclerites. **D: Calyx**, and un-retracted polyp head with sclerites in a **collaret and points** arrangement (ca. 3 mm). **E ᴀɴᴅ F: Contractile** polyps (E: *Klyxum*, F: *Efflatounaria*, 1 & 2 cm), **G: Calyx:** sclerites arise as a mound around the **retracted** polyp (here: *Astrogorgia*, 5 mm). **H:** Tall calyces within which lie the polyps of a *Briareum* sp. (1.5 cm). **I:** The tall, contracted polyp (**no calyx**) of an *Acanthogorgia* sp. (ca. 2.3 mm). **J:** Cross-section through a **scleraxonian**, showing the **cortex** and the central **medulla** (here: *Iciligorgia sp.*, 7 mm). **K:** The axis of Melithaeidae containing sclerites, segmented into **nodes** and **internodes** (3 cm, inlay: 2 mm). **L:** A holaxonian gorgonian showing the axis with its **hollow, cross-chambered, central core** (here: the plexaurid *Euplexaura*, 5 mm). **M:** The branch of a **calcaxonian** gorgonian (here: the ellisellid *Junceella*, 0.5 cm). **N:** Segmented axis in the family Isididae, with calcareous **internodes** and **nodes** made of gorgonin (here: *Isis*, 2 cm). Abbreviations: Ax = axis, Ca = calyx, Ce.co. = central core, Cn = coenenchyme, Co = collaret sclerites, Cor = cortex, In = internode, Lo = loculus, Me = medulla, No = node, Po = sclerites forming one of the eight points, Pol = polyp, Ri.cn. = ring of boundary canals, Sc.cn. = sclerites in coenenchyme. *Photos: A, C, E & F: KF, all others: PA*

Clavulariidae we find the next level of complexity, where vegetative production of multiple polyps from stolons produces polyps of similar height, which are only connected by canals running through the stolons (Fig. 8A). We also find tall, axial polyps that produce secondary polyps from their sides (Fig. 8B). In more advanced octocorals, the anthosteles (the lower, tubular extensions of the polyps) are embedded in the common, more or less fleshy, coenenchymal mass of the colony (Fig. 8C). In these more massive, erect forms it is often possible to divide the colony into recognisable parts such as the **stalk, stem, branches, lobes, disc** and **capitulum**. In the rod-, fan-, or bush-shaped gorgonians, the coenenchyme is generally thin, and covers a relatively hard, flexible, internal axis (Fig. 8D). Soft corals intergrade with these types of gorgonians through a number of more or less arborescent forms that have a central concentration of sclerites, loose, or bound with horn-like material.

Although some octocorals do not have **sclerites**, most possess them. They are found in greatly varying concentrations embedded in the coenenchyme, providing support and protection. Sclerites are produced by specialised cells, and are made of polycrystalline aggregates of calcite (a type of calcium carbonate) deposited over a fine, horny, proteinaceous, fibrous matrix. The shapes, sizes and colouration of sclerites vary widely among species. For example: up to 3 mm long spindle-shaped sclerites have been found in the surface of branches of the soft coral *Dendronephthya spp.* (Fig. 9A); 5 mm x 2 mm polygonal plates in species of the gorgonian genus *Paracis*; 1 mm spheres in a species of the soft coral *Capnella*; 0.5 mm scales with a projecting spike in species of the gorgonian *Echinomuricea*; and minute oval corpuscles, 0.02 mm long, in the species of the soft coral genus *Xenia* (Fig. 9B). Sclerites are the most important feature used in the identification of octocorals.

Most **soft corals** lack a rigid internal skeleton for support, instead water is pumped through the mouths of the polyps into the canal system, and the resulting internal hydrostatic pressure supports the colonies like a regular skeleton. Such temporary "scaffolding", which is known as a **hydroskeleton**, can be quickly dismantled in response to disturbance or severe wave stress: within less than a minute, colonies can excrete the excess water through the polyp mouths, and contract to small sizes. Lacking a rigid skeleton, most soft corals remain relatively small. There are, however, massive fleshy species with few sclerites that can grow to more than a meter thick, and even non-fleshy forms that for the most part are bags of water, which can stand as tall as two meters. The sclerites in soft corals are usually individually arranged in the colonial coenenchyme (Fig. 9A and B). However, in a number of genera, the sclerites are fused into clumps, or in tubes and volcano-shaped structures that house the polyps. The 'organ-pipe coral' *Tubipora musica*, is the most remarkable example of sclerite fusion in soft corals (Fig. 9C).

In some species of *Sinularia*, sclerites at the very base of the colony are very large and cemented together (Fig. 10), accreting to formidable "rock" and contributing to reef growth (Schuhmacher 1997). Such cemented layers of sclerites below flourishing colonies of *Sinularia* are commonly found on modern coastal reefs, but on such reefs this type of sclerite accretion has been going on for thousands of years (Konishi 1981, Kleypas 1996). The fossil rock is called **spiculite**, and probably represents the most common type of octocoral fossil. Although spiculites are known from the upper Tertiary (Russell Kelley, pers. comm.) these fossils are too young to provide information on the evolution of octocorals, but they give some insight into processes of reef formation. Spiculites were formed on turbid coastal fringing reefs of the Great Barrier Reef and other Indo-Pacific regions, and in this environment they appear to have contributed to a basis for modern reef growth after the end of the last Ice Age (Johnson and Risk 1987).

The arrangement of the sclerites in the colonies can be an important taxonomic feature. For example, in nephtheids groups of polyps may be supported by individual large sclerites, which are known as **supporting bundle** (Fig. 9A). Sclerites in the head of a polyp may be arranged in very characteristic way, called **collaret and points** (Fig. 9D). The collaret is the name given to a discrete series of long sclerites, often bow-shaped, that encircles the head of the polyp just below the base of the tentacles. The points are the eight groups of sclerites, often large and hockey-stick shaped, that sit above the collaret; one set of points is found on or below the base of each of the eight tentacles. Point sclerites can be present without a collaret. The arrangement of sclerites in a **calyx**, which is a sclerite-reinforced cylindrical or wart-like projecting mound around the polyp opening (FIG. 9F, G AND H), is also a taxonomic characteristic of some groups. Some tall, non-retractile polyps may appear similar to those bearing calyces (FIG. 9I), but closer inspection will allow to distinguish the two groups.

Most **gorgonians** have a relatively solid, internal, central axis over which a thin layer of coenenchyme and polyps grows. Although most species are much less than a meter tall, some colonies achieve remarkable size: it is not that uncommon on tropical reefs to find fan-shaped colonies two to three meters in diameter, and even larger forms exist in deeper and colder waters. The axes are built primarily of **gorgonin**, which is a hard, proteinaceous material similar to horn, and contains significant amounts of collagen or collagen-like substances. Many species incorporate large amounts of calcareous material. Species in one group of gorgonians (the **Scleraxonia** group) have axes that contain calcareous material in the form of sclerites. In the simplest constructions there is a central zone called the axial **medulla**, which is clearly or poorly separated from the outer **cortex** layer that contains the polyps (FIG. 9 J). This central medulla is composed of a dense aggregate of sclerites that are often a different shape to those in the outer cortex. It may or may not be extensively penetrated by canals, but it never has a hollow, central core. In its least complex construction, the sclerites are quite free and not bound together by

FIG 10: Sclerites in the base of *Sinularia* are very large, and often cemented together. **A:** Close-up of a colony base after the top was eaten by a cowry (*Ovula*). View: ca 3 cm. *Photo: Karen Gowlett-Holmes.* **B:** *Sinularia* trunks consisting of densely packed sclerites. This colony measures ca 1 m in height and 1.5 m diameter. Low Isles, Great Barrier Reef. *Photos: KF*

fibres of gorgonin (family Anthothelidae). These species can be broken quite easily and need relatively calm conditions in which to live. In its most complex form, the medulla sclerites are fused to their neighbours to form a meshwork that is completely embedded in a dense matrix of gorgonin (family Subergorgiidae). Such an axis is very tough. Somewhat in between we find the precious coral *Corallium*, where the axis is formed solely from sclerites that are cemented together by calcareous material to form a substance dense enough to be carved and polished. There is also a segmented axial construction that contains sclerites (family Melithaeidae, (FIG. 9K). The **nodes** from which branches arise are formed from loose sclerites embedded in a gorgonin matrix, and the straight segments in between, called **internodes,** are formed from fused and cemented sclerites.

Although gorgonians with an axial medulla consisting of more or less loose sclerites are relatively fragile, they still achieve remarkable sizes. *Paragorgia arborea*, which occurs in the north and south Atlantic, is the most noticeable. Colonies up to 3 m tall have been trawled from deep water, and basal fragments about 40 cm in diameter indicate that even larger colonies may exist.

Colonies of the same basic construction in which sclerites are arranged as a medulla and a cortex also occur in a thin mat-like, or membranous, growth form (Family **Briareidae**, FIG. 9H, and p. 154). Species that grow like this can hardly be referred to as gorgonians, but they are placed in the same categories as their arborescent cousins because of the similar construction. It is as though a thick branch from an arborescent colony has been slit to the centre, opened up and spread flat: the medulla becoming the basal part of the membrane, and the cortex becoming the upper part. Such species, which can encrust large areas of substrate, are further evidence that no separating boundary can be drawn between soft corals and gorgonians.

Two forms of construction, where the axis does not contain sclerites, exist among species that are perhaps commonly thought of as typical gorgonians. In one constructional form, the axis has a narrow, hollow, cross-chambered central core (suborder **Holaxonia**, FIG. 9L). In some cases this form of axis is permeated with non-scleritic calcareous material deposited in loculi (FIG. 9L), while in others it consists of pure gorgonin. The other form of construction does not have a hollow centre, being solid all the way through, and always permeated with large amounts of non-scleritic calcareous material (suborder **Calcaxonia**, FIG. 9M). There is also a very easy to identify variation of this latter axial form in which the calcareous substance, instead of permeating the gorgonin axis, occurs as solid segments that alternate with pure gorgonin segments (family Isididae, FIG. 9N). The nodes may be transparent, opaque, white or coloured, and are made of gorgonin, and the internodes are made of calcium carbonate. Although such a segmented axis cannot be flexed as far as a continuous axis before it breaks, individuals of several species with this construction are the largest gorgonians yet encountered. The sea-mounts in the deep waters of the Southern Tasman Sea have yielded incomplete, unbranched colonies 6 m tall, and huge, broken, branched fragments, like ivory logs in excess of 30 cm thick, from colonies whose original height we can only guess at.

The apparent dichotomy between branching gorgonians with an axis containing sclerites and those with an axis without sclerites is not as clear-cut as it would seem. There is one intermediate group of species (Family Keroeidae, p. 182) in which the axis consists of sclerites embedded in a matrix of gorgonin, which surrounds a hollow, cross-chambered, central core.

Reproduction and Propagation

Sexual Reproduction

In most octocorals, male and female reproductive structures are in separate male and female colonies. This mode of reproduction is called **gonochoric**. However, some soft corals, such as species of *Heteroxenia* and *Xenia*, are **hermaphroditic**, that is each mature colony contains both male and female reproductive structures (Benayahu and Loya 1984). Three types of reproduction occur in octocorals: (1) broadcasting of eggs and sperm, (2) internal brooding of larvae, and (3) external brooding of larvae (Benayahu 1991).

Alcyoniid soft corals and some gorgonians are **broadcasters**, that is they release large numbers of eggs and sperm into the water column where fertilisation occurs. Spawning is often synchronised by lunar phases and/or water temperature. This enhances the chance of fertilisation, as sex cells are receptive only for a short period of time, and are quickly dispersed by the currents. Larvae develop from the fertilised eggs, and remain planktonic for days to weeks, until they settle and transform

FIG. 11: Larvae of *Lobophytum compactum* under a confocal microscope. *Photo: Kirsten Michalek-Wagner*

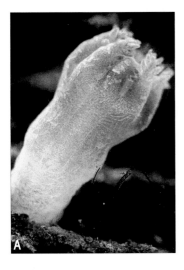  

Fig. 12: Early stages of recruitment and colony formation in the alcyoniid soft coral *Lobophytum compactum* from the Great Barrier Reef. A AND B: Founder polyps, C: initial stages of polyp budding. Views: 5 - 10 mm. *Photos: Kirsten Michalek-Wagner*

(**metamorphose**) into **founder** polyps – often tens to hundreds of kilometers away from their parents. Many gonochoric species, such as those of the genus *Clavularia*, the family Xeniidae and many gorgonians, are brooders. In this strategy, sperm, but never the eggs, are released into the water, generally a few hours after sunset. In **internal brooders**, a small number of eggs is fertilised and develop to larvae within the females. Days to weeks later, the larvae are released when they are almost ready for metamorphosis. Examples of internal brooders are *Xenia* and *Heteroxenia* (Benayahu et al. 1988, Zaslow and Benayahu 1996), and most gorgonians (Brazeau and Lasker 1990). In **external brooders**, (e.g., *Clavularia, Briareum, Rhytisma, Efflatounaria,* as well as some gorgonians), the fertilised eggs develop to larvae in mucus pouches on the surface of the mother colonies, where they remain until a later developmental stage. Brooded larvae are often negatively buoyant and may settle within meters of the mother colony.

After the planktonic or brooding stage, the oval, featureless planula larvae (FIG. 11) settle out of the water. Substrate types and light intensity determine their choice of settlement sites. Crustose coralline algae are preferred substrate for a number of species (Benayahu *et al.* 1989, Lasker and Kim 1996). Most larvae choose a consolidated hard substratum for attachment, whereas loose rubble or thick layers of sediments or turf algae are generally not suitable. Larvae often settle in little cracks or on the underside of small ledges. This provides protection against smothering by sediment or being scraped off by grazing fish or sea urchins, however for zooxanthellate taxa the cost of this protection is a lower light exposure and thus slower growth rates.

Once a larva is settled, it metamorphoses into a founder polyp (FIG. 12A AND B). It initially forms a short stalk at one end to attach to the substratum; then the other end flattens and eight tentacle buds develop around the mouth opening. After a few days, a complete octocoral polyp has developed. At this stage, some of the zooxanthellate species (see below) start taking in planktonic zooxanthellae through the mouth (in others, the oozytes are supplied with zooxanthellae prior to release into the water column; Benayahu *et al.* 1992). These algae are not digested, but are incorporated into the host tissue and start a new life as coral symbionts. Only few deep-water octocoral species remain permanently solitary, that is, each adult consists of only a single polyp. All other species develop from a solitary founder polyp to a colonial stage (FIG. 12C), consisting of a multitude of polyps, which are derived from the founder polyp by budding.

Asexual Propagation

Asexual propagation is common and often the predominant mode of reproduction in soft corals. It is
achieved by **runner formation**, colony **fragmentation**,
fission, or **budding**. Species of *Efflatounaria* form
runners (**stolons**) from the base of the colony, or extend
sterile branches to a length of 3 – 5 times the colony
size, which attach to the substratum (FIG. 13A). The
parent colony then translocates parts of its own body
mass through the stolon, and a new daughter colony is
formed which rapidly separates from the parent. The
stolon is later resorbed and disappears, resulting in two
independent and unconnected colonies of similar size.
Some species of *Sarcophyton*, *Lobophytum*, branching
Sinularia, *Nephthea*, and *Xenia* form vertical constrict-
ions through the colony, and eventually divide into two
independent smaller adult colonies (FIG. 13B). Other
species, such as *Sarcophyton gemmatum* or *Sinularia
flexibilis*, occasionally produce small buds on the edge
or base of the colony, which eventually drop off and
attach to the substratum as small separate colonies
(FIG. 13C). Species of *Dendronephthya* drop little polyp
bundles consisting of 5 – 10 polyps, which sink to the
bottom and attach by growing root-like structures (see
page 114; Dahan and Benayahu 1997). The gorgonian
sea whip *Junceella fragilis* has a similar strategy, dropping
its uppermost branch tip (hence the species name
fragilis!), which attaches to hard substratum on the sea
floor in the vicinity of the parent, and grows up as an
independent colony.

The reproductive strategies of different species are
often reflected in their extent of aggregation, and in
their ability to recolonise disturbed areas. Some
species of nephtheids and gorgonians rely largely on
larval settlement. They do not employ asexual
propagation, and tend to be fairy evenly distributed
within suitable habitats. The broadcasting of large
numbers of gametes, as found in the alcyoniids,
appears particularly advantageous for recolonising
remote habitats. The strategy of larval brooding and
effective asexual propagation, as found in many
xeniids, enables fast clonal growth, but slow
colonisation of remote locations. Species with fast
asexual propagation often grow highly aggregated,
in patches containing many genetically identical
clone mates.

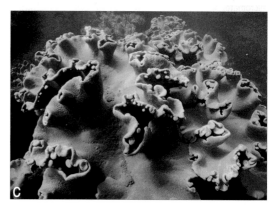

FIG. 13: Means of asexual propagation in octocorals. **A:** The
formation of stolons and small daughter polyps in
Efflatounaria. GBR, view ca 18 cm. **B:** Colony fission in
Sarcophyton, view ca 50 cm. **C:** Budding; here in form of
small, disk-like buds along the colony edge of a *Sarcophyton* from the GBR, colony size: ca 50 cm. *Photos: KF*

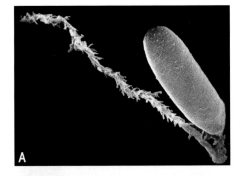

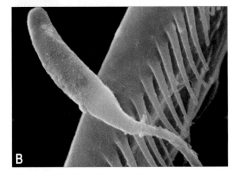

FIG 14. Nematocysts in octocorals, recorded by electron microscopy. **A:** Discharged nematocyst, showing the capsule and barbed tube. **B:** A discharged nematocyst (front). Its small size (ca 8 micrometer capsule length) is put in perspective when compared with the appendage of an artemia larva (back). *Photos: Richard Mariscal*

Nutrition

Feeding

The majority of octocorals are **suspension feeders** that filter small food particles from the water. Their diet consists predominantly of small (< 20 micrometer) particulate organic matter including phytoplankton cells, ciliates and other microzooplankton, and bacterioplankton (Fabricius *et al.* 1995a and b, Ribes *et al.* 1998, Yahel *et al.* 1999, Fabricius and Dommisse 2000). Larger particles that intercept with the tentacles or pinnules are captured, tested, and swallowed if suitable. The **nematocysts (stinging capsules)** of octocorals (BOX 5 AND FIG. 14) appear to be small and simple (Mariscal and Bigger 1977, Schmidt and Moraw 1982), so capture is restricted to small, weakly swimming or damaged zooplankton, such as early crustacean or bivalve larval stages (Lewis 1982, Fabricius *et al.* 1995, Ribes *et al.* 1998). Particles that are indigestible or too large (detrital or mucus flocs), or swim too actively (some zooplankton) are initially caught, but released after a few minutes.

The epidermis is densely covered with microvilli (Mariscal and Bigger 1977), which points at **absorption of dissolved organic matter** as food (Schlichter 1982a and b). However, concentrations of useable dissolved nutrients in most tropical waters tend to be low, and their role and contribution to the diet of octocorals are still not completely understood.

Food concentration and current speed affect rates of food intake and growth in soft corals (Fabricius *et al.* 1995a, b). Rates of food intake are highest at unidirectional, intermediate flow speeds (around 8 - 15 cm s^{-1}). At slow flow, rates of food transport to the polyps, thus food encounter and food intake are low. Very fast currents bend the polyps and afflict drag, which also lead to low rates of food intake. On the Great Barrier Reef, nearshore reef communities dominated by alcyoniid soft corals remove large quantities of suspended particulate matter from bypassing water (Fabricius and Dommisse 2000). They graze with greatest efficiency on small phytoplankton, but also make extensive use of other types of small particulate food.

BOX 5: Nematocysts:

A **nematocyst** (also called **cnida**) is a minute capsule inside a cell (**cnidocyte**), which is embedded in the epidermis. The capsule contains a fine, coiled up, hollow tube, which is attached to the mouth of the capsule. The capsule also possesses a lid and a small sensory hair-like trigger. Touch or chemicals can stimulate the trigger to discharge, and the tube instantly springs out, turning inside-out like the finger of a blown-up rubber glove, penetrating the object and depositing venom. Generally there are barbs on the thread-like tube (FIG. 14) which anchor the pierced object.

Octocoral nematocysts appear to be simply structured, small and ineffective compared with those of jellyfish, hydroids and sea anemones. They are unable to penetrate human skin; consequently soft corals and gorgonians cannot sting humans.

Photosynthesis

Food concentrations are often low in tropical waters, and many shallow water corals host symbiotic algae in their tissue to complement their diet (FIG. 15). These algae, known as **zooxanthellae**, belong to the group Dinoflagellata. In octocorals, zooxanthellae are embedded in gastrodermal cells, or apparently free in membrane-bound vacuoles within the gastrovascular cavities. Like any plant, these algae use sunlight, carbon dioxide, water and nutrients to produce sugars and cellular material. Unlike terrestrial plants however, these plants are embedded in animal tissue; sugars are passed on from the algae to the coral and contribute to its nourishment. In return, the coral supplies their algae with nutrients and carbon dioxide, and a protected microhabitat to live in.

Azooxanthellate octocorals (those which do not contain zooxanthellae) meet all their nutritional requirements by the intake of food. In contrast, **zooxanthellate octocorals** use photosynthesis to contribute to their energy needs, but they still need to obtain nutrients (nitrogen and phosphorus), trace elements (e.g., iron and vitamins), and additional energy by the intake of food. Rates of photosynthesis in zooxanthellate soft corals are often low compared to those in hard corals, and insufficient to even cover the basic respiratory carbon demand (Fabricius and Klumpp 1995). Such surprisingly low photosynthetic

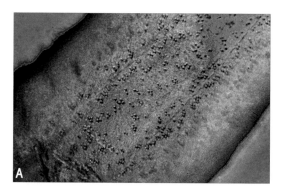

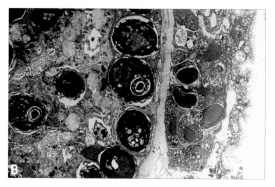

FIG 15, **A:** Confocal microscopy of the golden-brown zooxanthellae in the tissue of an alcyoniid polyp. *Photo: Kirsten Michalek-Wagner.* **B:** Transmission electron microscopy showing details of the dark, round zooxanthellae (6 micrometer diameter) in the gastrodermal tissue of a *Sarcophyton*. *Photo: Tina Tentori*

efficiency may be related to the ratio of the colony surface area to its volume. The photosynthetic rate depends on the surface area of a colony (a greater surface provides for greater light harvest), whereas its rate of respiration, representing its greatest energy demand, is a function of the biomass (a bigger body requires more oxygen and energy). Hard corals have a large surface area to volume ratio, as a very thin layer of tissue covers a massive, light-reflecting calcareous skeleton. Therefore, their photosynthetic efficiency is high, and respiration is low. The reverse is true for soft corals: the surface area is small in relation to the large biomass of the massive colonies, thus photosynthetic yields are relatively low, whereas respiration rates are relatively high.

Colony expansion during daylight increases rates of photosynthesis by as much as 30% (Fabricius and Klumpp 1995). Zooxanthellate gorgonians of the Florida Keys in the Caribbean are generally expanded and feed during the day, but many contract at night. However, on the Great Barrier Reef, a fair proportion of octocorals are contracted during the day, whereas almost all are expanded and actively feeding at night, and contracted at dawn and dusk (K. Fabricius, unpublished observations). The reasons for the periodic patterns of colony contraction and expansion are far from being understood. Day-time contraction seems to be triggered by slack or too strong currents, too high levels of irradiance, and disturbance by day-active polyp-eating fish, with trigger levels varying among species.

Ecology of Octocorals

Patterns of Distribution

Like plants, soft corals and gorgonians are subjected to the environmental conditions in which they have settled as larvae. They depend on a set of basic physical environmental parameters such as light, water motion, or sedimentation, however, the preferences and tolerance ranges vary greatly among individual species and genera (summarised in the "Soft Coral Atlas" of the Great Barrier Reef, Fabricius and De'ath 2000). Many genera inhabit only narrow bands along gradients of continuously changing environmental conditions (Fabricius and De'ath 1997, Fabricius 1998). Two environmental gradients are particularly well known: the depth gradient, and distance to the land. A range of physical environmental variables, such as wave exposure, light availability, turbidity, or exposure to terrestrial run-off, change along these two gradients, and it is often difficult to discern which of the many correlated variables ultimately control abundances.

On coral reefs on the edge of continental shelves such as the outer-shelf reefs of the Great Barrier Reef, on mid-ocean atolls, and desert-surrounded deep water basins such as the Northern Red Sea, the physical environment is characterised by very clear water with low particle loads, little resuspension of sediments from the deep sea floor, absence of disturbances related to terrestrial run-off, and occasional upwelling of nutrient-rich deeper waters. In such clear-water habitats, zooxanthellate species (in particular xeniids, nephtheids, alcyoniids, and *Isis hippuris*) are highly abundant (FIG. 16A, Benayahu and Loya 1981, Tursch and Tursch 1982, Dinesen 1983, Dai 1990, Fabricius 1997), and can cover more than 50% of terraces at around 20 m depth. Communities in shallower areas such as reef crests appear to be limited by too harsh wave energies in particular on the windward side, and yield low abundances and species numbers of soft corals (FIG. 18). Only a few small-growing colonies such as *Xenia* or *Paralemnalia* occupy shallow reef crests and flats just behind the wave-breaking zone. This area is not suitable for taller taxa such as *Lemnalia*, *Nephthea*, and most gorgonians. Gorgonians are restricted to current-exposed but wave-protected back-reef or deep front-reef environments.

Coastal reefs such as the in-shore reefs of the Great Barrier Reef (FIG. 16C) exhibit a wide range of habitat types, changing along gradients of turbidity, siltation, distance to the coast, and distance to rivers. Abundances of many species change along these environmental gradients. Reefs in very turbid water closest to the coast, where visibility may be less than 2 m, have very low abundances and species numbers of zooxanthellate soft corals. Their shallow areas may be dominated by *Briareum* and a few colonies of *Sinularia*, whereas a range of azooxanthellate gorgonians may be found at the reef base. At slightly greater distance to the coast, where visibility is typically around 2 - 5 m, assemblages may be composed of *Clavularia* and *Briareum*, together with some *Sinularia, Sarcophyton,* and *Klyxum*. Rich coastal alcyoniid assemblages, joined by a few nephtheids and xeniids, can dominate reefs in areas where visibility tends to be around 5 to 8 m (Fabricius and De'ath 2000). These assemblages can cover up to 70% of available space in flow-exposed and wave-protected, shallow areas of coastal reefs, leaving little space for hard corals.

The reefs on the middle of the continental shelf have the greatest diversity of octocoral species, although the cover is relatively low (often less than 5%; FIG. 16B). Distribution ranges both of inshore and offshore species overlap here, and, additionally, a range of typical mid-shelf species exist which are not found elsewhere.

In most parts of the world, the species richness of coral reefs increase towards the equator. On the Great Barrier Reef, the area of greatest octocoral biodiversity is the mid-shelf region between 11° and 13° latitude (FIG. 17).

FIG 16, A: Oceanic reefs, such as this reefs on the outer edge of the Great Barrier Reef, are often dominated by members of the family Xeniidae.

B: Octocoral communities on mid-shelf reefs of the Great Barrier Reef are highly diverse, with no single family being dominant.

C: Inshore communities on the Great Barrier Reef are often dominated by alcyoniid communities, as well as *Briareum*, *Clavularia*, and ellisellid gorgonians. *All Photos: KF*

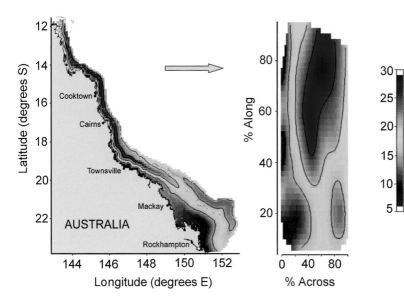

Fig 17: Biodiversity (generic richness) of octocorals on the Great Barrier Reef. The left panel in the latitude-longitude system has been converted into a coordinate system of relative distance across and along the continental shelf. Colours represent the total number of genera encountered in ten 15-min surveys per reef, assessed between 18 m depth and surface (from: Fabricius and De'ath, 2001).

The Physical Environment

Storms and Waves

Octocorals are susceptible to abrasion, dislodgment and other forms of damage by storm waves and the associated movement of sand and rubble. Windward crests of exposed outer-shelf reefs experience strong wave action on an almost daily basis, and few octocorals are found in this wave-exposed zone, which is the domain of stout hard coral species of the family Acroporidae (Fig. 18). Only thick-encrusting octocorals such as *Sinularia* or *Cladiella*, and small compact species of, for example, *Capnella*, *Paralemnalia*, *Asterospicularia* and *Xenia*, tolerate some wave action, and can be found in protected pockets on crests or on reef flats behind the breaker zone. Most other octocorals occur in highest abundances in less wave-exposed areas. Gorgonians in particular are predominantly found on wave-protected steeper slopes with strong currents. Many nephtheid species prevent tear damage during storms by temporarily contracting their colonies to a small proportion of their size, but nevertheless they are rarely found in very wave-exposed areas.

Extreme tropical cyclones, typhoons or hurricanes create a mosaic of disturbances, in which relatively small areas (at a scale of hundreds to thousands of square meters) tend to be completely destroyed. Close-by to such completely flattened

Fig 18: Very few soft corals are found on wave-beaten fronts of outer-shelf reefs. Great Barrier Reef. *Photo: KF*

areas may be patches where mortality is species-specific, with higher survival of robust species than of fragile species. Other, more protected areas on the same reef may show only little damage after a severe storm.

Currents

Current-swept and wave-protected areas, such as channels between reefs or islands, or flanks and ridges on deeper reef slopes, are often extremely rich in soft corals and gorgonians. Many octocorals require consistent and moderately strong, preferably unidirectional, currents for maximum food encounter. Currents transport food to, and waste away from colonies, and stimulate photosynthesis. Azooxanthellate gorgonians usually grow with their fans perpendicular to the predominant current direction to maximise food particle interception. In zooxanthellate species, growth forms sometimes represent a compromise between maximising light harvest (i.e., horizontal spread) and flow exposure (i.e., vertical extension).

Light

The distribution ranges of octocorals varies with light exposure which depends on depth, water clarity, and steepness and aspect of the reef slope. Particles suspended in the water not only make the water turbid but also absorb light, and a turbid site at 10 m depth can appear dark even in the middle of the day. Turbidity is greatest on shallow continental shelfs close to the coast or close to river mouths, where waves and tidal currents resuspend sediments and mud from the sea floor (FIG. 19).

Light promotes zooxanthellate species (irradiance levels directly affect rates of photosynthesis) and inhibits many azooxanthellate species (their larvae tend to settle in relative darkness). Depth limits of species increase with increasing water clarity: in turbid waters, zooxanthellate soft corals are restricted to the upper 10 m, whereas azooxanthellate species (in particular ellisellid, subergorgiid and plexaurid gorgonians, and *Dendronephthya*) predominate below 10 m depth (Fabricius and De'ath 2000, 2001). In clear water, the depth limits of some zooxanthellate species (in particular some *Sinularia*) extend far below 40 m depth, and many azooxanthellate species start to occur only below 25 m. Light exposure is also low on steep slopes and vertical walls, favouring taxa which tolerate low light such as *Briareum*, *Sinularia*, and azooxanthellate species.

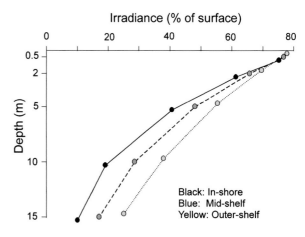

FIG 19: Light profile measured on 17 inner-, mid- and outer-shelf reefs of the central Great Barrier Reef (K. Fabricius, unpublished data).

Nutrients

Particulate and dissolved nutrients originate from very different sources in oceanic and coastal waters. Oceanic waters in the tropics are chronically nutrient-depleted by planktonic growth and associated grazing processes, hence they are clear and blue. However, currents sporadically push cool, nutrient-enriched, deeper water up the slopes of continents and atolls. Such **upwelling** significantly boosts the concentrations of available dissolved nutrients in offshore surface waters, although frequencies and extent vary widely between regions. Such regions are often dominated by xeniid soft corals (Fabricius and De'ath 2000), which have been shown to take up and use dissolved nutrients (Schlichter 1982 a, b). In shallow coastal waters, food concentrations are more variable and often higher than offshore, because of the **terrestrial run-off** of nutrients, and the **resuspension** of bottom sediments. In wet tropical areas, nutrient- and sediment-enriched river waters flood into the sea, enhancing concentrations of nutrients and sediments, and stimulating phytoplankton blooms and microorganism growth. Such areas tend to be dominated by alcyoniid soft corals in the shallow water (irradiance becomes limiting at greater depth). In many parts of the tropics, water pollution is considered a major disturbance for coastal reefs. Chronic discharges from sewage outfalls, and run-off from deforested and agriculturally used land add nutrients, top soil, and pesticides to the coastal waters. Run-off is said to contribute to the decline of nearshore ecosystems, and depending on the extent of run-off, problems have been recorded from the shore to tens of kilometers off the coast. Most zooxanthellate soft corals are missing in highly polluted areas, and those azooxanthellate gorgonians which are able to grow in polluted areas, often show high susceptibility to fungal infections, colonisation with algae, barnacles, bryozoans or anemones, and a high level of partial mortality (FIG. 20).

Sedimentation

There is significant natural variation in sedimentation on reefs, and tolerance to sedimentation varies widely among octocoral species. Enhanced sedimentation is suspected to have detrimental effects on the health of coral reefs, especially if it occurs in combination with enhanced nutrient concentrations, as is often the case in areas exposed to agricultural run-off. Thick sediment deposits can smother small colonies and coral recruits by restricting gas exchange between colonies and the water column. Sediments also negatively affect rates of photosynthesis (Riegel and Branch 1995), due to stress and due to the light absorption by the particles deposited on the colonies or suspended in the water. Many species, in particular most members of the Xeniidae and Nephtheidae, are restricted to relatively clear waters, whereas most members of the Alcyoniidae (in particular species of *Sinularia, Sarcophyton, Lobophytum,* and *Klyxum*) grow fastest and largest in moderately turbid yet well-lit coastal areas (Fabricius and De'ath 2000, 2001). In contrast, azooxanthellate gorgonians (e.g., many members of the families Ellisellidae and Subergorgiidae) are particularly abundant in some highly turbid areas even in shallow water, possibly due to the darkness associated with high turbidity and heavy sedimentation.

In the Caribbean, the fungus *Aspergillus*, which is typically associated with terrestrial soil and does not sporulate in sea water (Smith *et al.* 1996), causes widespread infections and mass-mortalities in *Gorgonia ventalina* and *G. flabellum*. The mass mortalities were linked to large river floods importing high sediment loads (Garzón-Ferreira and Zea 1992), and chronic infections were discussed to be a consequence of increased sedimentation from soil erosion (Nagelkerken *et al.* 1997).

Salinity

Not much is known about differences among species regarding their tolerance to fresh water. A salinity of around 35 parts per thousands (ppt) is normal for many Indo-Pacific environments, but values of 45

FIG 20: Partial mortality of gorgonians is said to be particularly common in polluted areas. Here: a colony of *Echinogorgia* growing in the waters of Hong Kong. *Photo: KF*

ppt are reported from the Northern Red Sea and the Arabian Gulf. Increased salinity due to evaporation is generally not a problem, as peaks rarely exceed more than 2 – 3 ppt above normal. However, reduced salinity on reef flats during monsoonal rain or in freshwater lenses of river plumes can be detrimental. In the Indo-Pacific, salinity below 30 ppt appears detrimental in particular for many xeniids, and values below 25 ppt are lethal for most species.

Temperature

The global distribution of hard and soft corals which host zooxanthellae is usually restricted to warm water regions (FIG. 22, page 28). Distribution limits in zooxanthellate species are caused by a narrow temperature tolerance of the symbiotic algae. Seawater temperatures dropping much below 18° C for extended periods of time are uninhabitable for most zooxanthellate species. The upper temperature limit of corals varies between regions, but is commonly around 31° C. A few regions which regularly experience summer maximum temperatures of 35° C (e.g., the Persian Gulf, or localised areas such as reef flats on wind-protected bays around islands) still support zooxanthellate corals, however often only a small number of species are found in such conditions. Only octocorals without zooxanthellae are able to grow in temperate or cold and deep waters.

Temperatures rising by only one or two degrees above the local long-term averaged summer maximum lead to the expulsion of zooxanthellae by photosynthetic organisms. Due to damage to the photosynthetic apparatus and subsequent loss of the brownish algae (Jones et al. 1998), zooxanthellate organisms including hard corals, soft corals, and clams turn pale or white, a process which is called **bleaching** (FIG. 21). Warm water appears to be the main cause for bleaching, but it can also be induced by other stress factors such as bacterial infections, fresh water, burial under sediment, or high irradiance of ultraviolet or visible light. Zooxanthellae populations in the colonies re-establish if the stress ceases after a short period. However if the stress is severe or persists for too long, the organisms eventually die.

Soft coral populations have been severely reduced in many parts of the tropics as a consequence of the mass bleaching event in 1998, but chances for survival vary among species. Bleached colonies of the family Xeniidae appear to be more likely to die than Nephtheidae, which in turn are more sensitive than many Alcyoniidae (Fabricius 1999). Records of bleaching may be biased towards greater numbers of records of bleached alcyoniids, as a result of their persistence: bleached species of the Alcyoniidae may survive for weeks to months and score higher in counts than bleached species of Xeniidae, which appear to die and disappear within a few days. Little information is available about bleaching sensitivities in other octocoral families.

The frequency of severe coral mass bleaching events has been increasing in the last few decades, and has been related to the global climate change, induced by deforestation and the burning of fossil fuels. Increased concentrations of carbon dioxide in the atmosphere cause a "greenhouse effect", resulting in a gradual increase of average and maximum summer temperatures in the tropics. Bleaching sets in when normal sea surface temperature summer maxima are exceeded by a few degrees for a few days. This is considered one of the most severe threats to the global health of coral reefs, and at present it is not known whether corals will be able to acclimatise to the warming of the tropical oceans (Hoegh-Guldberg 1999).

FIG 21: Reef communities experiencing mass bleaching and subsequent loss of coral cover after periods of unusually high sea water temperatures; here: coastal reefs of the central Great Barrier Reef in early 1998: **A:** A reef community dominated by alcyoniid soft corals at 5 m water depth. **B:** A large *Sarcophyton* colony at 10 m depth. **C:** A large dying colony of *Sinularia*. **D:** Many bleached alcyoniids survive for many months, while their colonies shrink to small sizes and undergo fragmentation. Here: a *Sinularia flexibilis* after two months in completely bleached state. **E:** Dying *Sinularia* leaving behind smooth, round mounds of sclerite rock ("spiculite"). **F:** the soft coral *Xenia* and the hard coral *Acropora* are both very susceptible to mortality after bleaching. *All Photos: KF*

Natural and Human-Induced Disturbances

The health of many coral reefs around the world is declining due to human activities (Wilkinson 2000). The three causes of most widespread damage are destructive fishing practices, global warming, and terrestrial run-off. Natural disturbances such as storms, flood plumes, or outbreaks of predators kill coral reef organisms, but on their own they are not known to cause detrimental long-term effects on reefs. On the contrary, they open the opportunity for less competitive species to re-establish by larval recruitment on to a disturbed area or by lateral growth of surviving colonies. If a reef is frequently disturbed (for example by frequent mass bleaching events, chronic pollution or destructive fishing), only fast-colonising "weedy" species will be able to occur, and species diversity will decline as result.

Worldwide increasing fishing pressure affects sea floor organisms directly due to damage caused by destructive fishing with explosives, cyanide, and bottom trawling. Removing fish predators causes additional indirect effects. The Caribbean and East Africa are well-documented cases where the removal of predatory and herbivorous fish has resulted in shifts from coral to macro-algae dominated communities (Done 1992). Such shifts often are not gradual, but may occur as a result of a single severe natural disturbance event in an area chronically deteriorated by human use, sometimes years after the onset of environmental deterioration (Hughes and Connell 1999). This is because adult hard and soft corals can survive greater pressure than their recruits, so adults may not die due to the deteriorated conditions, but may suffer physiological stress, partial mortality or reduced reproductive output

Fig. 22: Map of the distribution range of coral reefs in the Indo-Pacific Ocean and the Red Sea. *Source: World Conservation Monitoring Centre (UNEP WCMC).*

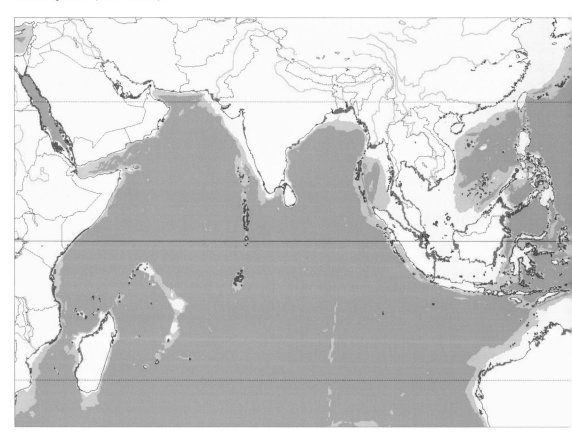

(e.g., Michalek-Wagner and Willis 2001), whereas the survival of the larvae or the settled coral recruits may be impaired. If recolonisation of a stressed site by coral recruits fails, the freed space may be occupied by other organisms such as seaweeds.

Some authors speculated that soft corals might outcompete and replace hard coral communities in disturbed reef environments. However, few reports and anecdotal observations exist to support this hypothesis (reviewed in Fabricius 1998). On clear-water reefs of the Great Barrier Reef, soft coral abundances did not appear to increase after the selective removal of hard corals by the coral-eating crown-of-thorns starfish *Acanthaster planci* (Fabricius 1997). Instead, the freed space remained empty (covered only by turf algae) even 8 years after the mass predation event, until fast-growing hard corals eventually recolonised the reef. Highly turbidity areas are marginal habitats for most zooxanthellate octocorals, and no observations exist of octocorals replacing hard coral communities in such environments. Most records of high soft coral abundances after hard coral mortality came from moderately productive, shallow and well-lit, wave-protected but current-exposed environments (Fabricius 1998). In such environments, some soft coral species (in particular *Sinularia flexibilis*, *S. cf. polydactyla*, *Briareum*, and *Clavularia*) are able to monopolise space over hundreds to thousands of square meters (Fabricius 1998). It appears that soft coral abundances are predominantly controlled by the physical environment (especially sedimentation, light availability, wave and flow exposure and steepness of the reef slope), thus space monopolies are restricted to physically well-defined, very narrow types of reef habitats.

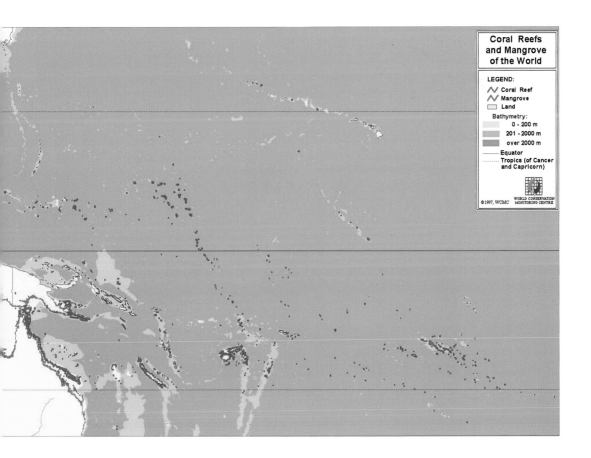

Chemical Ecology: Protection and Defence

Because soft corals and gorgonians are sessile (i.e., during their adult life phase they grow attached to the bottom, like plants), they cannot hide or run away from predators, space competitors or exposure to ultraviolet radiation. Instead, they produce a range of chemical substances, which effectively deter predators, prevent overgrowth by neighbours or fouling organisms, or screen out harmful radiation. Traditionally these substances have been called **secondary metabolites**, however they are by no means by-products. Their production requires specific enzymes, which are quite costly to produce and store. A change in terminology to **fundamental** and **complementary** metabolites has been proposed to replace the terms "primary" and "secondary" metabolites, in recognition of the physiological and ecological importance of the latter for the survival of these animals (Sammarco and Coll 1996).

Anti-Feeding Substances

Substances that deter predators are either toxic, or simply reduce palatability. Most belong to a group of chemicals called **terpenoids**, and are produced and stored by both larvae and adult corals of a wide range of species. Some species, such as the Red Sea *Dendronephthya hemprichi*, have low levels of chemical protection against predation, and have to compensate for high losses from grazers (such as many snails including *Ovula* and *Drupella*, the sea urchin *Diadema*, and butterfly fishes), by fast rates of growth and reproduction.

Anti-feeding substances are effective against most predators, but some specialist predators have genetically evolved the ability to break down particular toxins by metabolic processes. Probably the most well-known alcyonacean predator is the egg cowry (*Ovula ovum*) which grazes on the surface of colonies (FIG. 23). Massive scars from larger predators are commonly found on alcyoniids, and occasionally on *Efflatounaria* (FIG. 24). Scars on other species are either less common, less conspicuous, heal faster, or species are more sensitive so they die from the injury and leave no remains as evidence of the event. Some damselfish (e.g., *Neoglyphidodon melas*) and butterflyfish (eg., *Chaetodon melanotus, C. fasciatus, C. paucifasciatus*) feed on octocorals by selectively picking individual polyps, without causing significant damage to the colonies. Nudibranchs from a variety of families (e.g., Arminidae: *Dermatobranchus*; Dendronotidae: *Tritonia* and *Tritonopsilla*; and Aeolididae: *Phyllodesmium*) specialise in feeding exclusively on soft corals and gorgonians, and often imitate colouration and shape of the prey or host. These snails without shells not only digest the octocoral tissue, but also incorporate the animal's zoo-xanthellae into their own body for continued energy-producing photosynthesis. Additionally, they may incorporate the prey pigments to acquire the octocoral's colour for improved camouflage. The coral eating crown-of-thorns starfish *Acanthaster planci*, which removed up to 95% of hard corals during their epidemic population outbreaks on Indo-Pacific coral reefs, rarely eats octocorals (De'ath and Moran 1998), so outbreaks of this starfish have no direct effects on their abundance (Fabricius 1997).

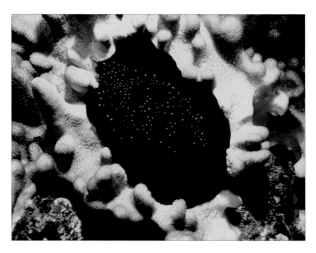

FIG. 23: A soft coral (*Cladiella*) eaten by the egg cowry, *Ovula ovum*. Kenia. *Photo: Mark Wunsch.*

FIG. 24: *Efflatounaria*, with several lobes bitten off by an unknown predator. Great Barrier Reef. *Photo: KF*

The ability of octocorals to heal wounds and regenerate lost branches due to predation, is of interest to the aquarium industry, as it enables fast and successful cloning of many species. Branches may simply be cut off existing colonies, and attached onto new substrate where they grow to complete colonies.

Interactions with Other Species: Symbionts and Epibionts

Because of, or despite, their chemical shield, octocorals are used as temporary shelters or permanent hosts by a large number of animals. The most conspicuous associates are barnacles, mussels, and feather stars which perch on gorgonians for better exposure to currents. Some cling fish, sea horses, brittle stars, ctenophores, snails, worms, shrimps and other crustaceans, including microscopic copepods, are also regularly found associated with octocorals. Such associations include a wide range of types, from seeking shelter to grazing on alcyonacean mucus or feeding on the coral tissue, and such epibionts are often well camouflaged, mimicking the colouration and surface patterns of their host. Even epiphytic algae can be found associating with octocorals, being commonly found attached to the horny perisarc on the bases of the stalks of soft corals. Often it is impossible to determine whether associates are mutualistic symbionts (both partners profit, e.g., the coral-zooxanthellae symbiosis), commensals (the host supplies food but doesn't experience significant damage or disadvantage), or parasites (the host experiences damage, ranging from minor to fatal).

FIG 25: A *Sinularia* overgrown by filamentous algae. Great Barrier Reef. *Photo: KF*

Anti-Fouling Substances

Anti-fouling substances prevent the growth of algae, fungi, bacteria, and other potential settlers on the colonies. If a colony loses (temporarily or permanently) its anti-fouling protection, it tends to be infected by fouling organisms within days. The loss of anti-fouling substances often appears to be related to severe environmental stress such as water pollution, or high temperatures. Stressed or weakened colonies cannot divert energy away from essential metabolic processes to the energy-expensive production of anti-fouling substances. In particular, alcyoniids and gorgonians are found with algal infections (FIG. 25). Gorgonians often do not recover from such infections, and dead branches, overgrown by algae and other organisms, may stay in place for months to years (FIG. 20). In contrast, some of the more persistent alcyoniids appear able to build up their protective chemical shield again, and recover within weeks to months. Some octocoral species occasionally produce a film of waxy mucus, which they use to cleanse the colony surface from fouling organisms and sediments. This mucus film is rapidly colonised by algae, which initially may be mistaken for an algal infection of the colonies. However, within days, the film is washed away by the currents and leaves behind a clean colony.

Allelopathic Substances

Probably around a thousand species of octocorals and hard corals coexist on the coral reefs of the Australian Great Barrier Reef. As habitat space is often limited, the sessile, perennial, and often long-lived species have to compete for space with their near-neighbours (FIG. 26), which is an energy-expensive process. In octocorals, the most commonly used strategy to inhibit growth and survival of neighbours is the slow release of chemical substances into the water (Sammarco *et al.* 1983, Maida *et al.* 1995). For example, a substance called flexibilide, released by *Sinularia flexibilis* at very low concentrations, causes tissue death in surrounding hard corals (Coll *et al.* 1982, Coll 1992). The settlement success of hard coral larvae is greatly reduced in a radius of up to more than half a meter around large *Sinularia* and *Sarcophyton* colonies because of their release of chemicals (Maida *et al.* 1995). Such chemical defence against space competitors, which is also found in plants, is called **allelopathy**.

Hard corals use a range of different strategies to inhibit neighbours, including the extension of elongated sweeper tentacles with high densities of stinging capsules in the bulbous tips, over-topping (depriving neighbours of light and water supply), and overgrowth. The Caribbean octocoral *Erythropodium caribaeorum* is the only octocoral recorded to employ **sweeper tentacles** (Sebens and Miles 1988); otherwise, the stinging capsules of octocorals are small and ineffective (FIG. 14). Both soft and hard corals are able to **overgrow** each other, but the outcome of such competitive interactions appears to be species- and habitat-specific, and clear competitive hierarchies are rarely established. Instead, competitive stand-offs, where both neighbours stop growing at the edge of contact, are the most common outcome of interaction. Most soft corals are too small to **overtop** hard coral neighbours.

FIG. 26: *Sinularia* overgrowing live *Acropora* colonies. Great Barrier Reef. *Photo: KF*

Internal Sunscreens

In response to the extremely high exposure to ultraviolet radiation found in clear tropical waters (Fig. 27), many shallow-water marine organisms, including soft corals, have evolved a group of unusual compounds, collectively called mycosporine-like amino acids (MAAs). Each of these MAAs absorbs ultraviolet radiation at a slightly different wavelength, so that together they function as a broadband filter of ultraviolet light.

Complementary Metabolites Used in Reproduction

The success of sexual reproduction in octocorals depends in many ways on complementary metabolites. Specific antimicrobial, anti-viral, and anti-predation substances (e.g., one with the name pukalide in *Sinularia polydactyla*) are contained in the eggs (Coll *et al.* 1989, Slattery *et al.* 1999). Pukalide also stimulates polyp contraction, thus facilitating the release of the eggs during spawning (Pass *et al.* 1989). Other substances in the eggs act as highly specific sperm-attractants (e.g., a substance called epi-thunbergol in *Lobophytum compactum*; Coll *et al.* 1995). The eggs are also supplied with high levels of MAAs, which protect the hatching planktonic larvae against damage by ultraviolet light.

Chemical Taxonomy

Many species produce a set of unique substances. This observation prompted the idea that relationships between certain species, especially those with inconspicuous or highly variable morphological features, may be determined by analysing their complementary metabolites. The discipline is know as **chemical taxonomy**, and may complement genetic and morphological analyses of relationships within some families. However, production, concentrations and composition of compounds are affected by environmental conditions, and variations of compounds within and among species are yet to be determined. It is unlikely that chemical taxonomy will ever replace more traditional taxonomic techniques, because of such variability, and the costs involved in biochemical analyses.

Bioprospecting

Pharmacologists and biochemists searching for new products such as pharmaceutics or anti-fouling paints are investing considerable resources in extracting the bioactive chemicals produced by octocorals. The natural bioproducts of a range of octocorals (e.g., some *Sarcophyton* and *Eleutherobia* species) are being trialed to test their pharmacological value. Others have already been proven successful. For example, the rapidly re-growing branch tips of the Caribbean gorgonian *Pseudopterogorgia elisabethae* are harvested on a commercial scale for their anti-inflammatory compounds which are used in some skin care products. This industry is reported to provide greater returns to local communities than previous (often dynamite-based) fishing. Such industry has the potential to contribute to the sustainable use of coral reefs, although unfortunately, some irresponsible bioprospectors have nearly wiped out local populations of desired invertebrates.

FIG. 27: Light levels can be exceedingly high on shallow reef-flats. Outer Great Barrier Reef, 2 m water depth. *Photo: KF*

Techniques for Examining and Identifying an Octocoral Colony

Species growing attached to coral reefs come in a confusing variety of forms. Coral reefs host 27 or so phyla, many of which are sessile. This makes it essential to understand how octocorals differ from other major groups of animals. One of the first macro-scale characteristics used to distinguish octocorals is the presence of polyps bearing eight tentacles, which nearly always have conspicuous rows of pinnules on both sides (SEE FIG. 2). Of course, polyps may be invisible because they are retracted, in which case it may be possible to make out the very evenly sized calyces or pores into which the polyps have withdrawn. Pores (oscula) in sponges and ascidians are often irregular in size and density (FIG. 28A), and are never associated with polyps, excepting a few rare cases of symbiotic association between sponges and hydroids. It should also be pointed out that there are a number of species of octocoral that become overgrown by a sponge without apparent detriment, and a few encrusting sponges and ascidians which resemble octocorals that often keep their polyps retracted during the day, such as *Briareum* or *Rhytisma*.

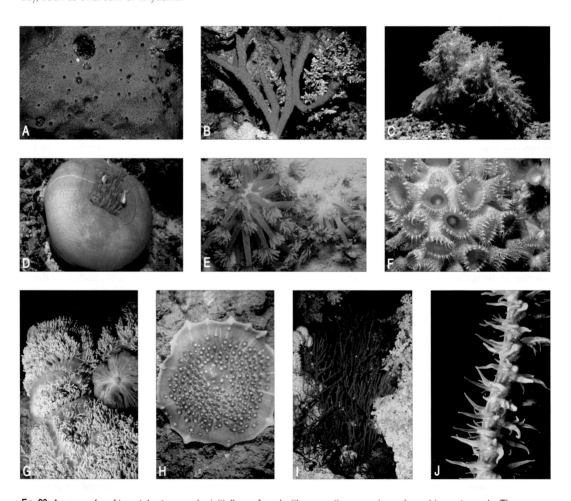

FIG. 28: A range of reef invertebrates may be initially confused with encrusting, massive or branching octocorals. Those are: A: an encrusting sponge, B: a branching sponge, C: the branching and stinging anemone *Actinodendron* which imitates the soft coral *Nephthea*, D: other anemones (here: *Heteractis magnifica*), E: the hard coral *Goniopora* sp., F: a zoanthid (*Palythoa* sp.); G AND H: two species of corallimorpharians, I AND J: the black corals *Antipathes* and *Cirripathes*. Photos: top right: PA, all others: KF.

Within the phylum Coelenterata, several types of animal have a soft and fleshy appearance just like the soft corals. The hard corals *Goniopora* and *Alveopora* (FIG 28E) are often mistaken for soft corals such as *Sarcophyton* when the skeleton is invisible under a thick layer of expanded polyps. Similarities to certain soft corals also exist in some anemones (in particular **Actinodendron** which looks just like the soft coral *Nephthea* but can inflict a painful sting; FIG. 28C) and zoanthids (FIG. 28F). Zoanthids that have colonised a gorgonian axis can look just like calyces, and can easily fool an observer. Further, black corals and hydroids may resemble some gorgonians and whip corals FIG. 28I AND J). However in all these cases, the presence or absence of polyps with pinnules, or the number of tentacles make it easy to distinguish the groups. There are a number of excellent field guides, textbooks and coffee table books (for example, Allen and Steene 1994, Gosliner *et al.* 1996) that can be used as an introduction to the different major phyla and classes of marine animals, as well as for the identification of common species.

With increased familiarity, most octocorals can be reliably identified to family level, and many to genus, using external features underwater. Characteristics that need to be taken into account are:
* the shape, typical size, hardness, softness, smoothness, prickliness, and colour of colonies;
* the arrangement, relative density, retractability or contractibility and roughness of the polyps;
* whether there are just autozooids, or siphonozooids as well (FIG. 5 AND 6);
* sclerite characteristics, such as presence of calyces (FIG. 9) or large sclerites in the colony base (FIG. 10);
* the presence of a solid axis, and whether it is segmented (FIG. 9); and
* characteristics such as response to touch, or smell of a specimen.

There is potential for error if just relying on underwater photographs to identify octocorals. For this reason our reference guide provides descriptive text and illustrations of the microscopic features needed to verify field diagnoses. The most valuable features are the sclerites, but they cannot be used in isolation, and their details need to be considered in conjunction with many other characteristics of any colony. In most genera that include more than just a few species, sclerite shapes can vary considerably. Space limitations prevent us from comprehensively illustrating all sclerite forms found in any single genus, but we have attempted to provide examples of the most common sclerite types for each of the relevant regions of a colony.

In order to examine the finer features, it is necessary to collect the specimen or a sample of it. Octocorals do not have the same sclerites distributed throughout the colony. Soft corals, generally speaking, have different sclerites in the surface of the polyp region (lobes, branches), the interior of the polyp region, the surface of the base and the interior of the base. The polyps themselves have different sclerites again. For a full diagnosis, until a satisfactory level of field confidence is achieved, it is therefore necessary to collect a sample with all of these regions present. Most colonies can regenerate if only a part is collected, however with small colonies the whole specimen may have to be collected. With larger species this is neither necessary not practical. It is sufficient in broad, thickly encrusting colonies to remove a segment shaped somewhat like a pie-slice, which includes upper lobes (branches or ridges), the surface layer of the colony side down to the base, and the attached interior portions. If identification beyond the generic level is anticipated, it is particularly important to have sufficient upper material so that evidence of the particular growth form is preserved. These sampling recommendations cannot be overstated, as trying to identify a species from just a lobe is akin to trying to determine the name of a fish from just the fins.

Removing a branch from a gorgonian is sufficient to obtain surface and interior material, together with some polyps, for sclerite examination. However, as with the soft corals, more material may have to be collected if a species determination will be attempted. Sclerites can vary, sometimes dramatically, from one branch to another. Sclerites near the colony base and from the axis, are commonly different to those in the younger colony parts. The branching pattern is an extremely important diagnostic feature. For example a number of genera in the family Ellisellidae only differ in growth form from each other, so it is important to know this character. Growth forms should be documented by an underwater photograph (rather than collecting a large sample), and it has to be emphasised that the impact of collection should always be kept to a minimum, in particular for species that appear to be long-lived and slow-growing. Furthermore, the collection of large specimens raises the additional problem of finding containers of sufficient size and of a construction suitable for alcohol storage. Numerous specimens reliably tagged (FIG. 29), or individually placed in plastic bags with holes, can be kept together in a single large container, but this is generally more suitable for soft corals than gorgonians, because the former pack more easily. The alternative is air-drying. It is more difficult to work with dry samples, they are more prone to abrasion and breakage, and are susceptible to mould and insect attack, but there is often no other option. A compromise is to keep a portion of a large specimen in alcohol.

Colony samples can be used for sclerite examination while they are still fresh, but it is more common for such laboratory work to be performed on preserved material. The safest and commonest method for long-term storage of small to medium sized octocoral specimens is to place them directly into 70% ethanol (ethyl alcohol) in fresh water. They will last indefinitely in this medium, and can be easily used for identification (FIG. 30). Relatively inexpensive industrial grade ethanol, about 95-96% pure (called

Species:	Menella	Photo: Film 3, UW
Colour:	bright red, white polyps	Size: 50cm
Reef/Site:	Stanley Rf.	Locality: Channel
Site ID:	251	Depth: 10 m
Date:	30/9/98	GPS: 14.10.09 S, 14443.14 E
Ref. No.:	099816	Collected by: K. Fabricius, AIMS

FIG. 29: Example of a label displaying information about sampling location etc. Such label should be added to all samples immediately, as features change rapidly after preservation (see Fig. 30).

Special Methylated Spirit in Australia, and Industrial Methylated Spirit in the UK) is quite sufficient for this purpose. It can be diluted with seawater, but this will result in a milky precipitate as the salts come out of solution. Keeping specimens for the long-term involves two processes. The first is called **fixation**, it prevents bacterial action and stabilises proteins and other tissue elements, which begin to degrade immediately after death. Fixed tissue becomes opaque. The second is called **preservation**, and serves to maintain tissues in their fixed state indefinitely. Sometimes the same chemical solutions are used for both processes, like ethanol as recommended above.

There are occasional recommendations for initially fixing octocorals in dilute formalin. Specimens preserved in this way look better externally than those placed directly in ethanol because there is limited shrinking. However, there are inherent problems in this method besides those associated with just handling this highly toxic chemical. Octocoral sclerites consist of calcium carbonate, and as such they are dissolved by acids. The formaldehyde in formalin gradually undergoes oxidation to formic acid, and there are many old and recent specimens sitting on museum and university shelves that testify to long term contact with this acidic medium. Any sclerites that remain in these samples look chalky white, they crumble easily, and under a microscope it can be seen that the tubercles, spines and ends have been corroded away. In more extreme cases, all sclerites will have disappeared completely. Sclerites are generally very small, and their surface ornamentation even finer, so relatively small amounts of acid etching can cause dramatic changes making identification of the specimens impossible.

Recipes for fixing octocorals in formalin can differ considerably, with recommendations varying from about a 4% to 20% solution. The most common is 10% formalin. The terms **formalin** and **formaldehyde** are often confused. As purchased, concentrated industrial strength formalin is a saturated solution of formaldehyde gas in water. This solution contains slightly less than 40% formaldehyde. A 10% formalin solution made up from 1 part of this concentrated solution and 9 parts of fresh (or distilled) water is actually 4% formaldehyde. In the undiluted form, the concentrate has an acidic pH of 3. Diluting to 10% with fresh water will bring this to a pH of about 6, which over time will become increasingly acidic due to oxidation. Most recipes advocate using buffered formalin to prevent the problems of acidity, despite the fact that in its buffered state its fixation effect is weakened. Buffering is the addition of chemicals to maintain a solution's pH in the event of dilution or addition of acids or alkalis, and it is important to use suitable reagents. Buffering a formalin solution with an acidic buffer will, of course, retain the acidity and not prevent sclerite corrosion. For those who insist on fixing octocoral material with formalin, what is really needed is neutralisation of the formalin combined with buffering to maintain this neutral state. As an attempt to achieve this, it is common to see recommendations for diluting concentrated formalin with seawater. Making 10% formalin from a concentrate at pH 3 using seawater will result in a mildly acidic or perhaps neutral solution, but it offers little buffering potential and will gradually become more acidic with time. A far better buffering method is as follows: use four grams of monobasic sodium phosphate monohydrate and 6 grams of dibasic sodium phosphate anhydrate per litre of 10% formalin (**not** 10% formaldehyde!).

The problems of dealing with the specimens after they have been fixed in formalin still remain. Recipes commonly say that the specimens should be fixed in buffered formalin for 24 or 48 hours, rinsed, and then transferred to 70% ethanol for preservation. Unfortunately the formalin that has penetrated the tissues will still be there, and even in the alcohol it will continue to oxidise to formic acid. Soaking specimens in tap water for long periods to remove the formaldehyde residue is also suggested, but no tests that have been done to establish how much time it takes to remove all traces. Fixing for short periods of 24 to 48 hours is really only useful for small specimens as formalin has a relatively slow penetration rate (about 5 mm per 48 hours in mammalian tissue). Large, fleshy specimens are unlikely to be fixed all the way through unless left in the formalin for perhaps 1 to 2 weeks, which leads to increased problems of removing all traces of formaldehyde before

Fig 30: Many colonies change size, shape, colouration, and surface structures once they are preserved. It is therefore always a good idea to make notes or take photographic records of features of the live specimens. *Photos:* **A:** *Carolina Bastidas,* **B:** *PA*

transferring the material to alcohol. Short-term immersion in formalin may not fix the deeper layers, which will continue to break down during any soaking out period in water. We therefore advise caution when using this chemical. If it is used, there should be regular changes of the storage alcohol until all traces of formaldehyde have gone. Formalin, even neutrally buffered, is likely to become acidic eventually, and should never be used for long term storage of octocorals. It is also very important to note in this age of increasing research at the molecular level, that contact with formalin renders specimens useless for DNA analysis.

Formalin fixation is recommended, however, if histological investigations are planned, and in order to optimise cellular detail there are a number of chemicals that can be added to the fixative solution. Unless the specimens are to be embedded in hard epoxy resin, they must be decalcified before they can be sectioned, in which case any previous sclerite degradation is immaterial. Specimens preserved in ethanol are not suitable for fine histological study.

Unless specific action is taken, all specimens, dried or preserved in fluid, will be in the contracted or partly contracted state. Specimens that are relaxed and then **narcotised** prior to fixing will undoubtedly reveal features, predominantly of the polyps, that are difficult to see in contracted colonies. Indeed, specimens preserved this way can take on a dramatically different aspect when compared to the contracted version. Such narcotising procedures, however, have not been extensively investigated over a wide range of octocorals, and the process can be very frustrating, prone to failure, and time consuming. Recommended methods are the slow addition of such chemicals as magnesium sulphate, magnesium chloride, formalin (!), or menthol, until the animal fails to react to stimuli, or the placing of a specimen directly into isotonic sea water and magnesium chloride. The narcotising can take many hours: specimens tested by probing too early can withdraw their polyps, never to be seen again; specimens left too long will begin to disintegrate; and specimens successfully narcotised may end up full of crystalline deposits after fixation and preservation. The procedure is really only suitable for a land- or ship-based laboratory situation. Because of the logistical problems, nearly all of the descriptive literature is based on specimens that have not been relaxed prior to being fixed. In fact, a contracted specimen is often preferable to a relaxed one when trying to make a species determination, because it will bear greater resemblance to the scientific description.

The techniques for examining sclerite shape and arrangements are quite simple, but demand access to a low power stereo (or dissecting) microscope, and a high power compound microscope. **Tissue is removed** from the sclerites by the use of bleach, which renders the material soluble and leaves the sclerites untouched. **Bleach** is a solution of sodium hypochlorite, and that supplied by vendors of commercial cleaning products or swimming pool chemicals (10 - 13 % chlorine) is ideal, though may not be available in less than several litres. Household bleach is available in smaller quantities, but it is more dilute and takes longer to dissolve the organic material. Bleach stock retains its activity longer if kept in a refrigerator.

In general, the samples to be examined only need to be very small. In many cases, a sample less than 2 mm square is ample, and should be cut from the relevant part of the colony with a scalpel. Larger samples may only be required if the sclerites are large, and this can easily be estimated using a stereo microscope. The sample is placed into one or two drops of bleach on a microscope slide. Once the bubbles have ceased, the sclerites can be spread out by stirring, and the sample viewed with a compound microscope. Use of a cover slip will guard against high-power lenses being dipped into the corrosive solution. Preparations that are allowed to dry can be rehydrated with a drop or two of water.

Until a satisfactory level of familiarity is reached, care should be taken to sample each region with as little contamination as possible. For example, surface samples should be cut deep enough to incorporate the longer sclerites that may stand perpendicular, but the amount of interior sclerites included should be minimal. For

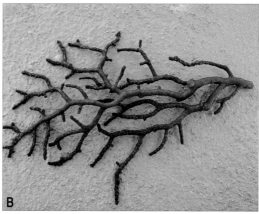

FIG 31: In large octocorals, it is generally enough to collect a small piece of a branch, if the colony size, shape, and colouration is recorded by underwater photographs or notes. **A:** some large fans of *Siphonogorgia* (Bali), **B:** a 10-cm sample of the same species from the GBR. *Photos:* **A:** *Mark Wunsch,* **B:** *KF*

anyone beginning this science, this will involve mainly guess work, but in this way the different types can be attributed to the correct layer. Similarly, calyces should be sampled separate from the surrounding areas, and retracted polyps carefully dug out for individual examination. Later, once the genera are better known, and the different sclerite forms become recognisable, this will no longer be so critical.

The arrangement of the sclerites in the polyps can be difficult to determine when the sclerites are not coloured. In this case it is often necessary to make the polyp tissue become transparent, known as **clearing**, before examining it under the microscope. One method is to place the polyp into phenol-xylene. This is a saturated solution of phenol crystals in xylene. A polyp can be placed directly from 70% ethanol into a drop or two of this mixture on a slide, but it will clear much quicker if it is washed in 95% ethanol first. Both chemicals, especially phenol, are toxic, and the mixture changes from colourless to red with time. Polyps cleared in this way harden, but can be recovered if necessary by washing with ethanol. A second method is to place a polyp into dilute bleach. When bleach is diluted the tissue disintegrates slowly without bubbles, and swells and clears as the process occurs. Microscope observations made as the transparent polyp gradually falls apart can reveal much extra data concerning the placement of sclerites. The dilution needed will depend on the chlorine level in the bleach, which becomes less with age. One drop of a solution containing 10% available chlorine and 3 drops of water is a good starting point.

Permanent microscope preparations of sclerites are made by placing washed and dried sclerites in a suitable mounting medium. It is advisable to begin with larger tissue samples than used for casual examinations. They should be placed in a small glass vial with sufficient bleach to dissolve all organic material. Once the bleach reaction has finished, the sclerites should be washed with a number of repeated changes of distilled water. The sclerites can be pipetted onto a microscope slide and left to dry, or allowed to dry in the vial and then tipped onto a slide. The mounting medium is then added, and a cover-slip applied. This washing and drying procedure is also the basis for preparing sclerites for electron microscope examination, but an extra step is advised. Once the reaction has finished, and the sclerites have settled to the bottom, most of the fluid should be removed with a pipette, and several drops of neutralised hydrogen peroxide should be added. The peroxide will cause violent effervescence that will shake off any bits of organic matter left adhering to the sclerites that would show up when using a scanning electron microscope.

Mounting media need to be acid-free, and have a refractive index substantially different from that of calcite. *Depex* is one medium that is in common use, but it suffers from the same problems that affect all evaporative-drying media when used with thick specimens: shrinkage. As the medium dries it loses volume and shrinks back under the edges of the cover-slip, initially requiring regular attention and topping up. One alternative method is to use an epoxy resin called ***Durcupan ACM***, made by FLUKA. It has the advantages of absolutely no shrinkage, and it will set hard overnight if heated. *Durcupan ACM* is generally marketed as a kit of 4 bottles, A/M epoxy resin, B hardener, C accelerator, and D plasticiser, but reagent D is not required for this method. Add 10 ml of reagent A to 10 ml of reagent B and mix thoroughly. To this add 0.4 ml of reagent C, mix thoroughly and put in the freezer. The bubbles generated during mixing will generally disappear overnight. In cold climates, briefly warming the combined reagents to 30-40°C will allow proper mixing. This resulting medium will last many months if frozen between uses. When removed from the freezer it must be warmed to lose its toffee-like consistency, and in warm climes this can be done by placing the bottle in tap water for a short while. Microscope slides made with *Durcupan ACM* will generally harden overnight if left in a 50-60°C environment. Some components of *Durcupan ACM* are toxic.

The identification of octocorals to species level is often extremely difficult, and in the diverse central Indo-Pacific fauna it is quite often impossible. There are large numbers of very similar species, many of which are still undescribed. Sometimes, taxonomic revisions are necessary before adequate identifications can be made, due to the poor state of the literature. In contrast, species identification may present less difficulties for well-described species in areas of low diversity, for example, along the northern and southern limits of coral reef distribution, as well as the westernmost and eastern regions of the Indo-Pacific. A considerable problem to people starting out in this field is that many of the existing taxonomic descriptions are fairly inaccessible, as they are scattered throughout the literature in small papers written in a number of different languages. Comprehensive treatises, as the renowned Monograph Series on hard corals by the Australian Institute of Marine Science (e.g., Veron and Pichon 1982), "Staghorn corals of the World" (Wallace 1999), or "Corals of the World" (Veron 2000), do not exist for octocorals, which is indicative of the relatively lower level of research that this group has attracted, and the problems still to be overcome. It is not an exaggeration to say that the majority of species named at this time need reviewing. The standard of the older literature commonly fails to make them recognisable, especially when they are considered against the huge numbers of specimens now available for comparison. Revisions only exist for a few major genera (e.g., *Sinularia*, *Sarcophyton* and *Lobophytum*: Verseveldt 1980, 1982, 1983), and a few others which are less frequently encountered in shallow tropical waters (eg. the family Ifalukellidae: Alderslade 1986, *Bellonella*, *Eleutherobia*, *Nidalia* and *Nidaliopsis*: Verseveldt and Bayer 1988, and *Minabea* and *Paraminabea*: Williams 1992b, Williams and Alderslade 1999). However, many of these and other genera will need to be revisited in the coming years. Species recognition is additionally hampered by numerous occasions in which diagnostic characteristics seem to overlap between species (e.g., the number of pinnules per tentacle in *Xenia*), and in which character variability within a species appears high (e.g., variability in growth form in some nephtheids).

Some of the apparent intermediate forms and within-species variability may be the product of **hybridisation**, which occurs when the reproductive products from parents belonging to different species combine, resulting in offspring that are viable and fertile. Colony shape and other characteristics of such offspring may show various degrees of intermixing between those of the parent species, and species boundaries become blurred for the researcher. Hybridisation is difficult, if not impossible, to detect in such primitive animals using colony shape and skeletal features. Three possible outcomes are: 1) new species are named based on hybrid specimens; 2) large degrees of variability are erroneously attributed to valid species; or 3) valid species are split into two or more nominal species. An intriguing hypothesis about the outcome of hybridisation is that species evolution, resulting from temporal or spatial isolation, may be reversed, so species that had previously split into separate species would come together again, mix their gene pools, and fuse back into a single species. This hypothesis, called reticulate evolution, is postulated and extensively discussed for hard corals (Veron 1995). Hopefully, techniques that look at enzymes and DNA will help to solve some of the above problems in octocorals.

Choice of Octocorals for a Coral Reef Aquarium

A tropical marine aquarium can be a truly fascinating addition to one's home. With care a number of species of invertebrates and fish prosper in an enclosed home aquarium system despite unavoidable differences from a natural living reef environment. Some zooxanthellate soft corals, zoanthids and corallimorpharians can be long-lasting and rewarding inhabitants, by contrast with hard corals which beginners find more difficult to keep in reef tanks.

Zooxanthellate soft corals have a beauty of their own, and some members of the families Alcyoniidae, Clavulariidae, and Briareidae, are hardy and have been kept alive, grown and propagated on a routine basis in aquaria for years (FIG. 32). Their suitability as aquarium specimens may be related to the fact that they are more tolerant of fluctuations in water quality in the field than other species. Some are so suited to mariculture that they can be grown on pebbles and sold in the aquarium trade rather than collected from the wild. Alcyoniid species, especially of the genera *Sarcophyton*, *Sinularia*, *Lobophytum*, *Klyxum*, and *Cladiella*, as well as *Clavularia* and *Briareum* are particularly recommended to novices or owners of aquaria with simple filtration systems. They naturally occur in the more turbid and nutrient-rich coastal environments and can sustain some fluctuations in salinity, nutrients and temperature. Some zooxanthellate nephtheids, including some species of *Capnella* and *Nephthea* are also easily kept and propagated. Slightly more demanding are the beautiful and tender *Xenia*, *Cespitularia*, and *Anthelia*, but again, tank propagation is possible in an aquarium with stable water quality. Experienced aquarists have also had good success in keeping zooxanthellate Caribbean gorgonians (Fosså and Nilsen 1996, Sprung and Delbeek 1997).

FIG 32: An 800 litre closed-system soft coral aquarium owned by Alf J. Nilsen, Norway. The soft corals grew so well that they had to be cut on a regular basis. The fungiid stony corals in the foreground also lived and grew for years and some spawned regularly. The tank is supplied with natural sea water collected off the coast of southern Norway, and water-quality is maintained mainly by protein skimming. Illumination is done by 3 x 250 watt metal halide lamps. Evaporated water is replaced regularly with carbonated water, while the pH is regulated by automatic addition of CO_2. *Photo: Bioquatic Photo : A.J. Nilsen*

Propagation in captivity takes advantage of the ability of many species to heal wounds and regenerate tissues. Branches or fragments are simply cut of the parent colony using sharp scissors or a scalpel. Freshly cut branches produce large amounts of mucus, and are susceptible to bacterial infection. Survival and attachment rates of the fragments are best in a corner of the tank where the water motion is sufficient to remove the mucus, and weak enough to leave the colonies in place. The specific weight of xeniids and some nephtheids is similar to that of water, so they are easily picked up and shifted by water flow. To avoid this, fragments may be kept in place with the aid of a tooth pick (inserted vertically into the base of the fragment, or pierced horizontally through thin branches), which can be tied or glued to the substrate or a piece of coral rubble. Alternatively, they are placed in a little crack in the substrate, and kept in place with a small metal-free ring of mesh. Tissues start to attach wherever they are in gentle contact with the substrate.

Aquarium suppliers harvest species of *Dendronephthya* and other azooxanthellate soft corals and gorgonians because of their beautiful bright colouration. Lacking symbiotic algae, these species are entirely dependent on suspension feeding of minute particles for their nutrition, because they do not photosynthesize (page 19). They constantly filter microscopic particles and absorb dissolved nutrients from the water, thereby processing large quantities of water every day. In the wild, many of these species feed continuously at night, and additionally during the day when currents are strong. The twin dietary requirements (continuous supply of small food items, including live, highly diluted phytoplankton; and dependency on strong currents) are hard to meet in tanks. Moreover, as is the case for most hard corals, azooxanthellate octocorals are very sensitive to changes in the water quality and physical conditions in the tank, are susceptible to infections and algal overgrowth, sensitive to strong light, and wither if currents are insufficient. At present, most of these species are almost impossible to grow or propagate in tanks. They are generally wild catches (i.e., collected as adults from the reef), and even very experienced aquarists seem unable to keep them alive for more than a few weeks. Due to their poor life expectancy in captivity, it seems irresponsible to attempt their husbandry, and for those gorgonians which are slow to grow and reproduce, harvesting from the wild is likely to be unsustainable.

The recommendation of choosing photosynthetic alcyoniids, nephtheids, clavulariids and *Briareum*, as opposed to azooxanthellate species is only a broad guideline, as differences in the requirements between species even within a genus can be large, and certain aquaria tend to be more favourable for "problem species" than others. The health and well-being of an aquarium and its inhabitants are fundamentally linked to the aquarist's knowledge about the organisms' biology. Well-designed water flow, ample aeration, filtration of particulate and dissolved nutrients, protein skimmers, sufficient light intensity with natural spectral composition, continuous fine-tuning of alkalinity levels, and a thermostat need to be understood in order to guarantee a healthy balanced system. In the past decade, great advances have been made in controlling water quality and understanding the natural requirements of marine organisms, and many excellent books cover in detail the maintenance of a reef aquarium at home. Two recently published sets of books present a wide range of issues on Indo-Pacific soft corals and Caribbean gorgonians, their ecology and husbandry, and give detailed information on general coral reef aquarium maintenance: these are Fosså and Nilsen: *The Modern Coral Reef Aquarium*, Volumes 1 - 3 (1996, 1998, 2000; Volume 4 in prep.; Volumes 1 - 6 available in German), and Sprung and Delbeek: *The Reef Aquarium*, Volumes 1 and 2 (1994, 1997). Those interested in marine aquaria should refer to such books, specialist aquarist journals, and even consider joining an aquarium society for assistance and advice before setting up a marine home aquarium and increasing the demand on wild stocks.

Surveying Octocoral Communities

Coral reefs are among the most diverse and species-rich ecosystems on earth. Despite constant change, coral reefs are clearly structured (Done 1982, Fabricius 1997, Devantier *et al.* 1998). To classify reefs by ecological and economic values requires comprehensive surveys and the establishment of biodiversity monitoring programs. So far, reef surveys have commonly focused on fish and hard corals, but valuable additional information is gained by also surveying octocorals and monitoring their changes. Octocorals may be particularly suitable proxies for reef habitat types, and may help to evaluate system responses to various types of reef degradation because:

- abundances of octocoral species are strongly determined by their physical environment and water quality (Fabricius and De'ath 1997 and 2000, De'ath and Fabricius, 2000); abundance and species composition in the communities may therefore serve as indicators for change. For example, the species richness of zooxanthellate octocorals declines where turbidity increases (Fabricius and De'ath 2001);
- many octocoral species appear to be long-lived, if environmental conditions do not change and no catastrophic disturbances affect the communities. Measurements on life histories of octocorals are sparse, however anecdotal observations indicate that the life expectancy in particular of some Alcyoniidae and gorgonians could be many decades, and growth rates are often low;
- octocorals have low levels of predation and grazing: by contrast to hard corals, fish and macro algae, abundances of octocorals appear to be little affected by predation or herbivory. There are no generalist predators known to affect octocorals to such an extent as the coral-eating crown-of-thorns starfish (*Acanthaster planci*) and the snail *Drupella* diminish hard corals;
- tropical octocorals show no signs of seasonality, thus surveys can be conducted at any time of the year.

Some excellent books introduce and evaluate the various methods of surveying coral reefs (e.g., English *et al.* 1999). There are many ways to conduct a survey. Quadrats are only recommended when very small and abundant organisms are targeted. Transect methods are generally considered superior to quadrats because of their better spatial representation. There are several types of transects to choose from, the most common ones being line intercept and belt transects. In line intercept transects, a tape measure is stretched over the substratum, and the location and type of all benthos intercepted by the tape is recorded. This results in good estimates of cover and abundances (Marsh *et al.* 1984). In belt transects, the number and location of all target biota are recorded within a strip of substrate (e.g., 50 cm wide). Using this method, additional ecological data (other than estimates of cover and abundances) can be gained, such as sizes, number of lesions, or measures of aggregation. In swim surveys, areas of many thousand

FIG 33: Taxonomic inventory of octocorals using swim surveys by the first author, here on an outer-shelf reef on the Great Barrier Reef. *Photo: Carolina Bastidas*

square meters are investigated, collecting data of estimated abundances and/or cover (FIG. 33; Miller and De'ath 1996, Devantier *et al.* 1998, Fabricius and De'ath 2001). Swim surveys are the most suitable technique if taxonomic inventories and rapid ecological assessments are to be conducted, because the large area covered guarantees superior representation of rare species. However depending on the design of the study, they are less sensitive in detecting subtle changes than the other techniques.

The following decisions have to be made about survey techniques and design prior to commencing a study:
- method of data recording: data can be recorded visually by a trained observer, or alternatively, the information can be recorded by photography or video;
- transect length and width;
- number of replicate transects per site;
- depth and orientation (slope-parallel or perpendicular) of the transects within a site;
- establishment of permanently marked transects for long-term studies, or unmarked randomly positioned transects.

The decisions will depend on habitat type, the type of information required (whether to establish cover or species inventories, or to determine changes in communities). Generally, more information is gained by sampling several depth zones rather than replicating within a given depth zone. If field time is limited, it is better to survey one transect each at 3 depth zones rather than 3 transects at one depth within a site. This is because many of the ecological processes vary greatly with depth (e.g., speed of recovery after crown-of-thorns outbreaks, intensity of storm damage, intensity of bleaching). On the Great Barrier Reef, octocoral communities change more between 3 and 18 meters depth at any one site, than between latitude 10° and 20° South at the same depth (Fabricius and De'ath 2001)! It is strongly recommended to refer to the technical literature before making decisions on the most appropriate survey and monitoring techniques before deciding on a study design (English *et al.* 1999). Here, we provide lists of alcyonacean genera which can be reliably distinguished in field surveys, and which contain ecologically meaningful information. TABLES 1 AND 2 represent examples of how a survey sheet for visual surveys and for video/photographic surveys may be structured. TABLE 1 lists the genera distinguishable in well-focused photographs or high-resolution videotapes taken from around 50 cm distance, and TABLE 2 lists genera identifiable in the field after a moderate level of training.

Most of the octocoral families on Indo-Pacific coral reefs have distinct ecological characteristics, and as a first step, much ecological information is gained by identifying octocorals to family level, instead of recording just "Percent soft coral cover". For example, Ellisellidae and other gorgonians without zooxanthellae, as well as many Clavulariidae and Briareidae, occur in highest densities in turbid water (Clavulariidae and *Briareum* preferring the shallow, and gorgonians the deeper areas). High abundances of Xeniidae and Asterospiculariidae, in combination with some Nephtheidae, are only found in clear water environments away from sources of terrestrial run-off. Alcyoniidae are found in greatest abundances in relatively well-lit, shallow and moderately productive coastal waters, although they can be present in a very wide range of ecotypes down to 30 m depth. The family Nephtheidae is more diverse in its ecological requirements, and habitat preferences vary widely between genera. In contrast to the xeniids and the alcyoniids, nephtheids are unsuitable as indicators of habitat types on family level, whereas the abundance of many neptheid genera can be related to the physical environment. Distribution ranges and ecological characteristics of the most commonly encountered families and genera are described in the "Soft coral atlas of the Great Barrier Reef" (Fabricius and De'ath 2000).

In general, significantly more ecological information is gained by conducting surveys on generic rather than familial level. The following chapters should provide the information necessary to carry out this task.

TABLES 1: Data sheet for photography or video analyses. Listed are the most common octocoral genera on the Great Barrier Reef that an experienced observer can reliably recognise on high-resolution images. Additional genera encountered are being recorded in the empty rows.

	Transect 1	Transect 2	Transect 3
Clavularia			
Carijoa			
Tubipora			
Sinularia			
S. flexibilis			
S. brassica			
S. …			
Dampia			
Cladiella			
Klyxum			
Rhytisma			
Sarcophyton			
Lobophytum			
Nephthea/Litophyton			
Stereonephthya			
Scleronephthya			
Dendronephthya			
Lemnalia			
Paralemmalia			
Capnella			

	Transect 1	Transect 2	Transect 3
Nidalia			
Siphonogorgia			
Chironephthya			
Nephthyigorgia			
Studeriotes			
Asterospicularia			
Xenia			
Efflatounaria			
Cespitularia			
Sympodium			
Sansibia			
Anthelia			

	Transect 1	Transect 2	Transect 3
Briareum			
Iciligorgia			
Solenocaulon			
Alertigorgia			
Subergorgia			
Annella			
Melithaeidae			
Acanthogorgiidae			
Plexauridae			
Rumphella			
Pinnigorgia			
Dichotella			
Junceella			
Ctenocella			
Plumigorgia			
Isis			
Unrecognised gorgonians			

TABLES 2: Data sheet for visual surveys. Listed are the most common octocoral genera on the Great Barrier Reef that an experienced observer can reliably recognise in the field. Additional genera encountered are being recorded in the empty rows.

Reef   Location   Date   Time in   Tr. Nr.

GPS lat and long: Exposure   Wind speed   Dist to shore

Depth (m)	18 - 13	13 - 8	8 - 3	3 - 1	1
Visibility (m)					
Slope angle					
Flowspeed					
Waves (0,1,2,3,4)					
Relief (0,1,2,3)					
Sediment (0,1,2,3)					
HC %					
SC %					
Rubble %					
Gravel %					
Sand %					
Silt %					
Turf algae %					
Coralline algae %					
Macro-algae %					
Sargassum %					
Dead coral %					

Remarks:

	18 - 13	13 - 8	8 - 3	3 - 1	1
Nephthea					
Stereonephthya					
Scleronephthya					
Dendronephthya					
Lemnalia					
Paralemnalia					
Capnella					
Sinularia					
S. flexibilis					
S. brassica					
S. ...					
Sarcophyton					
Lobophytum					
Cladiella					
Klyxum					
Rhytisma					
Heliopora					
Tubipora					
Clavularia					
Briareum					
Annella					
Subergorgia					
Melithaeidae					
Acanthogorgiidae					
Plexauridae					
Pinnigorgia					
Rumphella					
Dichotella					
Ctenocella					
Junceella					
Plumigorgia					
Isis hippuris					

Xenia/Heteroxenia	
Efflatounaria	
Cespitularia	
Anthelia	
Sansibia	
Sympodium	
Asterospicularia	

Although our knowledge of octocorals has increased greatly in the last half a century, notably aided by the advances in laboratory and underwater technologies, there is still much we need to know, and much we will probably never know due to the absence of a good fossil record of these animals. The latter is an obvious factor contributing to the deficiency of the knowledge needed for a satisfactory division of the Order Alcyonacea into suborders; there being consensus only about the holaxonian (see page 58) and the calcaxonian (page 62) gorgonians. Time has seen the other families subjected to various groupings, at various levels of classification, but newly discovered material, and the re-evaluation of old findings, has shown such overlap between these groups that their maintenance as suborders is difficult to justify. We include them here as "groups" only. The names are often useful when talking in generalities, but using them as discrete categories should be viewed with some caution.

THE *STOLONIFERA* GROUP

A few authors still group the following families of soft corals under the name Stolonifera. The name was originally proposed as an order to cover all those genera having species where the polyps arise separately from ribbon-like stolons, and are not united side by side in a common fleshy mass. However, a review of all the genera that grow in a manner similar to this shows there is a clear series of transitional forms between those where the polyps are not united, and those where they are united in a common coenenchyme. Stolons can merge to form membranes, membranes can grow quite thick, and "separate" polyps can show various amounts of fusion. The decision whether to categorise a particular genus as a stoloniferan becomes so subjective that the name plainly has limited classificatory value, except perhaps to indicate the growth form of some of the more primitive genera.

Family Clavulariidae

Colonies in this family are usually quite small, in some cases minute. They consist of cylindrical or bluntly conical polyps usually joined only at their bases by reticulating stolons, which may coalesce into thin membranous expansions. In some genera, tall cylindrical polyps develop long secondary polyps that resemble branches. In a few instances, polyps are also connected by extra, transverse, bar-like stolons above the basement layer. There is little overlap between the shallow-water Indo-West-Pacific genera in this family and those in other families. The only similar genus is *Cornularia*, but it grows in European waters, and has a thicker chitinous envelope around each polyp. This thick cuticle forms a theca-like cup into which the polyp can withdraw, and so *Cornularia* is not included in this family.

A thin, soft, brownish, transparent covering (called the **perisarc**), enveloping the stolons and the lower parts of the polyps, is characteristic for most genera in this family (it is worth noting that probably all soft corals have a perisarc coating near their very base, or at least between their base and the substrate upon which they are growing). However, the thin perisarc can easily be overlooked, and in many cases it is extremely difficult to detect without histological preparations. Sometimes, light reflecting from wrinkles, or the presence of detritus and foreign organisms adhering to the perisarc surface, is the only clue to its presence. Because there are many other morphological characters that can be used to distinguish the genera we have illustrated, searching for the perisarc is not necessary.

Nearly all of the species in this family have sclerites, and it is not unusual for these to be fused into clumps or tubes. Sclerites include such forms as smooth branched rods, and prickly or tuberculate 6-radiates, spindles and platelets.

Clavularia: p. 66

Cervera: p. 68

Stereosoma: p. 69

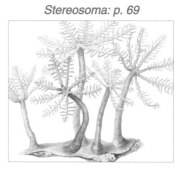

Carijoa: p. 70

Paratelesto: p. 72

Family Coelogorgiidae

Coelogorgia: p. 74

Coelogorgia is the only genus in this family. Colonies often appear as bushy clusters up to 1 m or more tall, and could be confused with gorgonians. The main stems and branches are actually the thick walls of axial polyps. The main axial polyp alternately buds off long, thin, daughter axial polyps that in turn bear very short lateral polyps. Branches can arise in one plane or spirally. Colonies are attached to the substrate by a complex stolonic holdfast. The walls of axial and lateral polyps contain tuberculate spindles.

Family Tubiporidae

Tubipora is the only genus in this family. Colonies are hemispherical and massive, to thick and encrusting, and consist of vertical, red, hard calcareous

Tubipora: p. 76

tubes connected at various levels by horizontal, stolonic platforms. Colonies can be up to 50 cm in diameter. Each tube is formed and occupied by a single polyp, which is connected to the other polyps in the colony by canals inside the horizontal plates. Stolons and non-extendable parts of the polyps are covered in a thin, soft perisarc.

The ALCYONIINA Group

Although the name Alcyoniina appears now and then in textbooks and sites on the world wide web, it is rarely seen in modern literature dealing with the identification of octocorals. The category is meant to include all those soft corals in which groups or all of the polyps are united side by side in a common, fleshy mass, without a supporting axis of horny and/or calcareous material. However, it is now obvious that a complete series of intermediate forms link the primitive soft corals in the Stolonifera group to a number of the genera included in the Alcyoniina and Scleraxonia group (page 53).

Family Alcyoniidae

Colonies in this family are the dominant reef dwelling octocorals in the Indo-West-Pacific. Their growth form is often termed "massive", referring to the fact that their dimensions are similar in all directions, and that the polyps are united to form fleshy masses. This is, of course, a generalisation, and there are species that grow in the shape of cigars or carrots, others form extensive encrusting mats several meters across and only a couple of centimetres thick, and one genus grows as membranous expansions only millimetres thick.

In most colonies, there is a bare basal section, the stalk or trunk, and an upper, polyp-baring part, divided into lobes, ridges or short branches. In some, the upper surface is flat, undulate, or loosely pleated around the margin. Massive or carrot-like colonies cut in half will reveal many very long canals inside, which are the gastric cavities of numerous primary autozooid polyps extending from the colony base to the uppermost regions. On the colony surface, siphonozooid polyps may be present amongst the autozooids.

Sclerites include tuberculate or prickly spindles, clubs, 6- or 8-radiates, ovals and dumbbells. The interior of a colony may be compressible and jelly-like if the sclerite content is low, and ridged and solid if it is high.

Sinularia: p. 78   Sinularia: p. 78   Dampia: p. 82

Cladiella: p. 84   Klyxum: p. 86   Rhytisma: p. 88

Sarcophyton: p. 90   *Lobophytum: p. 94*   *Lobophytum: p. 94*

Paraminabea: p. 98   *Eleutherobia: p. 100*   *Bellonella: p. 102*

Family Nephtheidae

An often bewildering set of genera are grouped in this family, some probably erroneously. Most can be described as bushy, globe-shaped or arborescent in growth form, while a few others are massive, and one genus consists of numerous finger-like lobes united by a common base. Many genera contain highly coloured species.

In most cases the polyps, singly or in small clusters, are more or less restricted to the upper and outer twigs or branches. In a few cases, polyps grow directly on main branches. If an arborescent colony is cut longitudinally, a small number of broad primary polyp canals can be seen, which split into groups that extend up into the distal lobes and branches. The canal walls are generally thin, with few sclerites, permitting colonies to easily inflate with water and dramatically increase their size.

Although some genera are soft and floppy, nephtheids are commonly known for their rough or often distinctly prickly feel. In the upper colony region, this is caused by the strong, protective sclerite armature of the polyps, which may project for several millimetres beyond the polyp head. In the stem and branches, the sandpaper-like texture can be attributed to numerous, strongly sculptured, spiny sclerites in the surface layer. Sclerite forms include prickly needles, leafy clubs, irregular shaped spiky forms, and tuberculate and thorny spindles, often extensively ornamented along one side.

Three genera (*Nephthea*, *Litophyton* and *Stereonephthya*) have been the source of much confusion for octocoral taxonomists. They contain a complex array of sclerite forms, of sclerite arrangements in the polyps, and of polyp distribution. Fortunately, a complete revision of this taxonomic chaos by our friend and colleague Leen van Ofwegen, of the Nationaal Natuurhistorisch Museum in the Netherlands, is nearing completion after years of research. The description of these genera has been kindly given to us by Leen, with the proviso that it is possible the details may be altered when the research is finalised, as there are still a number of problems to be solved.

Nephthea: p. 104

Litophyton: p. 106

Stereonephthya: p. 108

Scleronephthya: p. 110

Dendronephthya: p. 112

Umbellulifera: p. 116

Leptophyton: p. 118

Pacifiphyton: p. 119

Lemnalia: p. 120

Paralemnalia: p. 122

Capnella: p. 124

Capnella: p. 124

Family Nidaliidae

There are two groups combined in this family: one that includes genera with the shape of the soft corals, and another that includes forms that look much like gorgonians. In both groups, colonies are stiff, often brittle, and their outer walls are constructed of closely packed, large, tuberculate spindles, longitudinally arranged, and surrounding a group of large canals of the primary polyps. Those species that resemble gorgonians do not have a central axis, they are easily broken, and the interior sclerites are not organised into outer cortex and central medulla layers. The other group includes species which are small, unbranched and torch-like, or have small numbers of lobes or finger-like branches. In these, the outer wall of sclerites is relatively thin and surrounds a gelatinous interior, with fewer sclerites, and broad canals.

Polyps are always retractile, usually into prominent calyces. One deep-water West African genus (not covered by this book) has siphonozooids. Colonies may be highly coloured.

Nidalia: p. 126   Siphonogorgia: p. 128   Siphonogorgia: p. 128

Chironephthya: p. 130   Chironephthya: p. 130   Nephthyigorgia: p. 132

Family Paralcyoniidae

The upper part of a colony, which carries the polyps, is retractable into the lower part. In most genera the polyps are carried on lobes or branches, which when expanded can resemble colonies more likely to be found in the families Alcyoniidae or Nephtheidae. However, attached to rock or buried in the substrate, these colonies have a soft or hard, cup-like or capsule-like, base into which the lobes and branches can be withdrawn.

Sclerites are mostly tuberculate rods and spindles. *Carotalcyon*, a deep-water Japanese genus (not covered by this book), has sclerites that are mostly 8-radiate capstans, and is the only genus with siphonozooids in this family.

Studeriotes: p. 134   Studeriotes: p. 134

   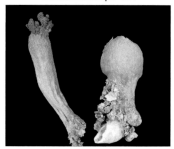

Asterospicularia: p. 136

Family Asterospiculariidae

Asterospicularia is the only genus in the family. Colonies are small, with a short stalk surmounted by several low rounded lobes of very regular shape and size (about 2 cm in diameter). The monomorphic polyps are confined to the lobes, and they are quite short. *Asterospicularia* is very similar in external features to species in the family Xeniidae, but has eight fully developed mesenteries.

Colonies are densely packed with minute stellate sclerites, like those found in didemnid ascidians.

Family Xeniidae

Xeniid colonies are mostly small and soft, and often quite slippery to touch because they can generate a lot of mucus when handled, which can make preservation difficult. Some species are also prone to autolysis shortly after collection, and can disintegrate into an unrecognisable mess. Growth forms include: thin, membranous expansions, upon which tall polyps may stand; short, cylindrical stalks, undivided or branched, terminated with domed polyp-baring regions; colonies that are multiply branched and delicate; and those with just a few short, blunt lobes. Two nominal genera (which may be the same) have siphonozooids present at least during some periods of their life, and a number of species have pulsating polyps, where the autozooids continually open and close the tentacle basket.

Nearly all of the species in this family have sclerites, and often lots of them. But one of the major features of this family is that the sclerites are not tubercular. Instead, they are nearly always minute platelets or corpuscle-like forms, with a surface that appears almost smooth at the magnifications of a light microscope. These sclerites often reflect light so as to appear opalescent, imparting brilliant blues and greens to live colonies. The genus *Anthelia* is an exception to the rule, having sclerites in the form of minute rods with a coarse crystalline surface.

Another major feature of the family is that only the asulcal, or dorsal, pair of mesenteries retain their filaments in the adult polyps (SEE FIG. 7 AND BOX 4, p. 10); the filaments on the other six being absent or rudimentary. Such detail, however, can only be observed in histological preparations.

Xenia: p. 138

Heteroxenia: p. 140

Heteroxenia: p. 140

Funginus: p. 142

Efflatounaria: p. 144

Efflatounaria: p. 144

Cespitularia: p. 146

Cespitularia: p. 146

Sympodium: p. 148

Sansibia: p. 150

Anthelia: p. 152

Anthelia: p. 152

THE *SCLERAXONIA* GROUP

The name Scleraxonia refers to colonies having inner axial-like layers or actual axial structures formed solely from, or containing, sclerites. The group primarily consists of those species in which the coenenchyme is divided into an outer layer, the cortex, where the polyps are situated, and an inner layer, the medulla (FIG. 9, p. 13). The medulla may be continuous or segmented, but it always contains sclerites, and never has a hollow, chambered core. The group, however, contains very diverse forms. Although most are upright and branched, others grow as encrusting membranes, and in a number of species the distinction between the cortex and medulla layers is very difficult to detect. These latter species are little different from many placed in the families Clavulariidae and Nidaliidae. Some authors treat the group as a suborder, but its seems likely the families have arisen from several separate evolutionary lines, and much more research is needed to establish which should be classified as a valid suborder.

Family Briareidae

Briareum (which incorporates *Solenopodium* and *Pachyclavularia*) is the only genus in the family. Most species in the family look like encrusting or lobular soft corals, but the family is placed alongside other families that contain genera with the aspect of gorgonians because of the arrangement of the coenenchyme into a cortex and a medulla. Different species of *Briareum* may occur as thin sheets encrusting rock, or dead or live substrate. They may grow as small clusters of knobs, as tall finger-like lobes, or as large tangles of cylindrical branches, which may be hollow. The polyp-bearing surface of the colony may be virtually smooth, or it may have calyces that can be up to 20 mm tall. Colonies are easily broken.

The basal layer, or medulla, which is attached to the substratum, or forms the centre of branches or hollow tubes, is deep magenta due to the colour of the sclerites. The upper layer, or cortex, may be magenta or almost white depending upon the amount of coloured sclerites it contains. The gastric cavities of the polyps are restricted to the cortex, and the gastrovascular canals penetrate the medulla, sometimes extensively, but there is no distinct zone of boundary canals separating the cortex from the medulla.

Except for the most basal layer of the medulla, the sclerites are all spindles, sometimes branched, with low or tall, spiny tubercles arranged in relatively distinct girdles. The most basal layer generally includes multiply-branched, reticulate and fused forms with very tall, complex tubercles.

Briareum: p. 154

Briareum: p. 154

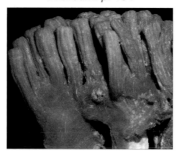

Briareum: p. 154

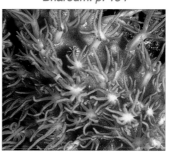

Family Anthothelidae

Colonies may occur as thin, encrusting sheets, or they may be upright and digitiform, or clavate, or richly branched. The branches may be cylindrical, they may be flattened and fusiform in cross-section with fine edges, they may be tubular, or they may be solid but have spatulate ends. An as yet undescribed species forms large flat fronds, looking like a sponge or an alga. Branches or sheets are divided into two layers, an outer cortex and an inner medulla, which are separated by a relatively distinct ring, or layer, of boundary canals. The gastric cavities of the polyps are restricted to the cortex, and, in the branches, the medulla is only rarely penetrated by the gastrovascular canals. Because there is no consolidated axis, colonies are easily broken. The sclerite forms of the cortex include tuberculate spindles, rods and 4-6-radiates. Those of the medulla include prickly needles and rods, smooth irregularly branched forms, and tuberculate radiates, spindles and ovals. In the encrusting colonies, the sclerites in the most basal zone may be fused. In most species, the sclerites in the cortex are usually different to those in the medulla, and may have a different colour.

Iciligorgia: p. 158

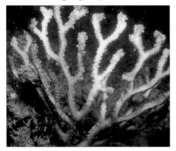

Solenocaulon: p. 160

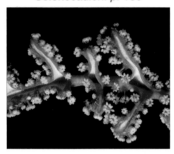

Alertigorgia: p. 162

Erythropodium: p. 164

Family Subergorgiidae

The medulla is relatively consolidated, making the colonies of this family difficult to break because the sclerites are branched, interlocked, partially fused, and also embedded in a tough matrix of gorgonin. There is a ring of boundary canals directly outside the medulla, below the cortex, which can be easily peeled off, but there are virtually no canals running through the medulla. Colonies grow either in an arborescent manner with free branches, or as netlike fans, and are never membranous. Although a genus recently found in the Antarctic includes tuberculate sclerites in the medulla, the axial sclerites are usually just smooth and sinuous, relatively long, and form a network. Sclerites of the cortex are predominantly tuberculate spindles and ovals, and small, irregularly shaped dumbbell-like sclerites often referred to as double-heads or double-wheels. Using just external features, it is not difficult to confuse species in this family with those in families with no sclerites in the axis.

Subergorgia: p. 166

Annella: p. 168

Family Melithaeidae

The most noticeable feature of this family is that the axial medulla is segmented, and consists of a series of short, rounded nodes, alternating with longer, narrower internodes (FIG. 9, p. 13). The nodes contain separable, smooth, cigar-shaped sclerites embedded in a gorgonin matrix, and are firm but compressible. The hard internodes are also formed from cigar-shaped sclerites, but all except the more recently formed sclerites are inseparably fused with calcite. The cortex is easily peeled from the axis, and canals penetrate the soft nodes. There are two other families where colonies have jointed axes, but only the Parisididae, which follows, has sclerites in the axis.

With few exceptions, branching occurs at the spongy gorgonin nodes, and produces both bushy, and broad, fan-like colonies. The swollen axial nodes on the stem and main branches are generally conspicuous, while those at the points of branching within the fan may not be so obvious. Colonies are quite fragile, breaking at node-internode joints.

Melithaea: p. 172

Mopsella: p. 174

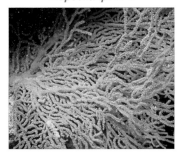

The sclerites are usually coloured, and besides the axial rods they include tuberculate clubs, ovals and spindles with ridges or spines along one side, leaf-clubs, and multirotulates (which look like two or more buns pressed together).

Wrightella: p. 176

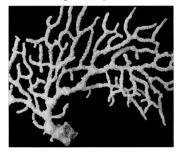

Clathraria: p. 177

Acabaria: p. 178

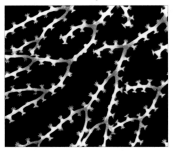

Family Parisididae

Parisis: p. 180

The axis of colonies in this family is composed of solid, longitudinally grooved, calcareous internodes formed from fused tuberculate sclerites, which alternate with horny nodes containing sclerites in the form of lobate rods. Branching emanates from the calcareous internodes.

There is only one genus, *Parisis*, which is predominantly found in deep water. The colonies form richly branched fans, and the sclerites of the cortex are mainly large, tuberculate plates or blocks, which form a pavement-like surface.

SUBORDER *HOLAXONIA*

The next few families make up a group called the Holaxonia. The name refers to the axis being 'whole', in the sense that it does not contain sclerites, however, there is a remarkable exception to this. The structural support in the Holaxonia looks like an axis in the more classical sense, and appears as a brown or black, hard but flexible skeleton (FIG. 9, p. 13). It is made up of layers of horny scleroproteinous gorgonin, alone, or together with various amounts of calcareous material. The gorgonin is present as concentric, tightly connected fibre bundles or lamellae. Between these are small spaces called loculi (singular: loculus), in which the calcareous material, if present, is laid down as amorphous or fine crystalline deposits. In the centre of the axis there is a relatively narrow, predominantly hollow, tubular space, which is partitioned into a long series of small chambers. This hollow structure is usually referred to as the cross-chambered central core (or cord), but it is sometimes called the axial medulla. The loculated gorgonin structure in which the core lies is sometimes called the axial cortex. The two together make up the axis proper, and are the equivalent of the medulla in the Scleraxonia group. The tissue covers the whole axis structure, and contains the sclerites and the polyps. This is the equivalent of the cortex layer in the Scleraxonia group, but is generally just referred to as the coenenchyme. To avoid confusion, the terms cortex and medulla (as used for genera in the Scleraxonia group) are rarely used when referring to the Holaxonia.

The exception to the absence of sclerites in the axis is the family Keroeididae. In this family the hollow, cross-chambered, central core is still present, but the calcareous material consists of sclerites, which are tightly or loosely bound in a gorgonin matrix, forming the bulk of the axis.

Other families where the colonies have a solid axis but no hollow central core are in the suborder Calcaxonia.

Family Keroeididae

This is the special group mentioned above, where the calcareous material in the axis is in sclerite form, as found in the Scleraxonia group. However, down the middle of the axis there is a hollow cross-chambered central core, and as such these animals bridge the gap between the Scleraxonia and the Holaxonia. The central core may be hollow, or it may be filled with very loosely bound sclerites. The surface sclerites of Indo-Pacific keroeidid gorgonians are commonly large spindles or plates that form a pavement-like surface. Such a structure makes it easy to confuse species in this family with some Plexauridae, which do not have sclerites in the axis.

Keroeides: p. 182

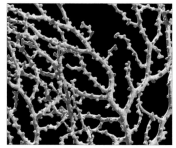

Family Acanthogorgiidae

Colonies in this family have a predominantly black, purely horny axis that is very difficult to cut, and has a wide, hollow, soft, cross-chambered central core. But the most noticeable feature is that the polyps, which are completely covered with straight and curved spindle-shaped sclerites, are not retractile, and are very conspicuous. These spindles are commonly arranged in eight double rows, forming chevrons (obliquely angled double-rows).

Colonies are richly branched, fan-like, net-like, bushy, or untidy and tangled. In some species, the coenenchyme is so thin that the black axis can be seen through it, but in others it is quite thick, full of sclerites and opaque. Sclerites are mostly tuberculate spindles, but tripods and capstans also occur.

Many species of the genus *Acanthogorgia* are easily recognised by the crown of sharp spines around the top of the polyp. Others are less spiny, but the thin, transparent coenenchyme is a clue to their identity. In other genera, however, the coenenchyme is thick, and the polyps can look like the calyces found in species from the family Plexauridae. Pulling apart a calyx to reveal a contracted polyp within can easily separate these forms.

Acanthogorgia: p. 184

Anthogorgia: p. 186

Muricella: p. 188

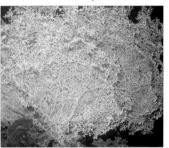

Family Plexauridae

Colonies in this family have a black to brown, horny axis with a wide, hollow, soft, cross-chambered central core, and numerous hollow spaces (loculi), that on rare occasions are filled with non-scleritic calcite.

Colony shapes resemble trees, bushes and fans (often net-like), and are sparsely to richly branched. Some species have massive holdfasts consisting predominantly of aragonite.

The polyps are retractile, often within prominent calyces, and are usually armed with large sclerites in a collaret and points arrangement. Sclerites are often quite large, longer than 0.3 mm, and some as long as 5 mm. They are tuberculate, sometimes thorny, and the tubercles are rarely arranged in regular whorls. Sclerites called thorn-scales often occur in the walls of calyces; these usually have a basal, spreading, root-like structure, and large, distal spines or blades that often protrude beyond the rim of the calyx.

The sclerites in the coenenchyme come in a very wide range of forms, including thorn-clubs, leaf-clubs, ovals, spindles, thornspindles, thornscales, plates, blocks, capstans, stars, and rosettes.

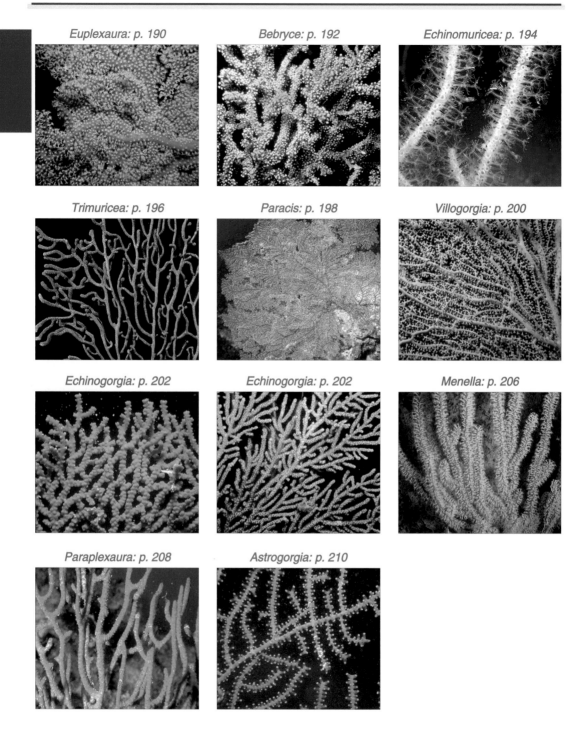

Euplexaura: p. 190

Bebryce: p. 192

Echinomuricea: p. 194

Trimuricea: p. 196

Paracis: p. 198

Villogorgia: p. 200

Echinogorgia: p. 202

Echinogorgia: p. 202

Menella: p. 206

Paraplexaura: p. 208

Astrogorgia: p. 210

Family Gorgoniidae

Colonies in this family have a black to brown, horny axis, but in contrast to the previous two genera, the hollow, soft, cross-chambered central core is usually narrow, and the axial cortex surrounding the core is very dense, with little or no loculation. Non-scleritic calcareous material in the form of carbonate hydroxylapatite may sometimes be present in the axis.

Colony shapes resemble trees, bushes and fans (sometimes net-like), and are sparsely to richly branched. In a few species the coenenchyme grows web-like between the branches, producing broad, flat fronds. In others, the branches are flattened like ribbons, or are tripod-like in cross section. Branching can also be pinnate. A few species have massive holdfasts containing calcite or aragonite, and sometimes large amounts of apatite.

Sclerites are generally small, usually less than 0.3 mm in length. The polyps may have no sclerites at all, but if present they are generally small, flattened rodlets with scalloped edges. The sclerites in the rest of the colony are nearly always spindles with the tubercles arranged in whorls. Some species have curved, asymmetrically developed spindles called **scaphoids**. Polyps are always retractile, sometimes into low calyx-like mounds. The sclerites in these mounds are never specially modified.

Rumphella: p. 214

Hicksonella: p. 216

Hicksonella: p. 216

Pseudopterogorgia: p. 218

Guaiagorgia: p. 220

Pinnigorgia: p. 222

SUBORDER *CALCAXONIA*

This is a newly established suborder, grouping those families in which the colonies have a solid axis without a central, hollow core (FIG. 9, p. 13). The name Calcaxonia alludes to the very large amounts of non-scleritic calcareous material present in the axis. This material, in the form of calcite or aragonite, is either deposited between the horny fibres, or is present as a central core or as solid internodal sections alternating with nodes of pure gorgonin in a segmented axis. The solid calcareous internodes can be white or coloured. In a continuous axis the colour can vary from black through to pale grey, depending upon the extent of the calcareous deposits. In some families the axis has a green to golden metallic or iridescent sheen.

Family Ellisellidae

The major feature of this family is the characteristic form of the sclerites. They are shaped like clubs, double heads and spindles with distinctly separate, papillate tubercles. There are also capstans with cone-like processes.

Colonies have a strongly calcified, continuous axis, and can be unbranched, loosely branched, or form broad, flat fans that may be net-like. It is very important to note the way a colony branches, because genera with the same sorts of sclerites are separated on growth form. Polyps are highly contractile, but not retractile. When contracted they may fold over and lie against the branch surface, or just close up to form a small mound, which may look like a calyx.

Ellisella: p. 224   *Viminella: p. 226*   *Ctenocella: p. 228*

Heliania: p. 230   *Junceella: p. 232*   *Dichotella: p. 234*

Verrucella: p. 236

Nicella: p. 238

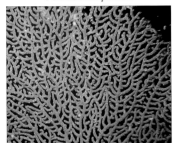

Family Ifalukellidae

The most notable feature of this family is the characteristic form of the sclerites. They are very small, mostly 0.025-0.09 mm, with a coarse crystalline structure, and present in various shapes including ovals, subspheroids, cylinders, peanut-shapes, crosses, multiradiate stars, angular rodlets, and irregular forms.

Ifalukella: p. 240

Plumigorgia: p. 242

Colonies have a planar or bushy growth form with pinnate or irregular lateral branching. The retractile polyps are extremely small, and the axis is highly calcified, continuous, and can form a large, strong calcareous holdfast.

Family Primnoidae

The major feature of colonies in this family is that the sclerites are virtually all scales or plates. Colonies can be unbranched, but are usually tree-like, bushy, bottlebrush, or fan-like, and richly ramified, often in a pinnate or dichotomous manner. The axis is continuous, and strongly calcified, generally grey to black in colour, and sometimes has a metallic sheen. Down the centre of the axix there is a solid core of calcareous material.

The polyps are always non-retractile, often arranged in pairs or whorls, sometimes fused side to side, and sometimes upside-down. They are protected by a covering of broad scales, often arranged in eight rows, and an operculum of eight triangular scales usually covers the in-folded tentacles when the polyps are contracted. The sclerites in the rest of the colony are also scales, sometimes narrow, and often modified as plates or blocks.

Plumarella: p. 244

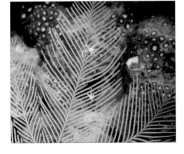

Sclerites in the form of scales are also found in the family Isididae and, to a lesser extent, in the Chrysogorgiidae. The Isididae are easily distinguishable because they have a segmented axis. In the Chrysogorgiidae, the sclerites are usually rods, sometimes flattened and sometimes broad and therefore becoming scale-like. Furthermore, most chrysogorgiids have very regular, geometric branching patterns, such as spirals, and are predominately restricted to deep water.

Family Chrysogorgiidae

Stephanogorgia, with branches growing in a zigzag manner, appears to be the only genus encountered at diving depth in the Indo-Pacific. Genera in this family include species characterised by their very repeated geometric growth patterns, such as neat spiral, dichotomous or uniserial branching; only a few species are unbranched. The other notable character of the family is the metallic sheen of the highly calcified axes, which varies from iridescent blue-green to gold. It is not unusual for the axis colour to be seen through the coenenchyme and the sclerites, which can form a very thin layer. In numerous species, the sclerites are virtually restricted to the polyps and there are few or no sclerites in the coenenchyme at all.

Stephanogorgia: p. 246

Sclerites are mostly rods, sharp and needle-like, or flattened and often broad and scale-like with a smooth or granular surface. The scales can have side branches and be quite ornate. There are also a couple of genera that have thick tuberculate scales and spindles.

The only other tropical gorgonian with a zigzag growth form is *Zignisis,* but it has a segmented axis.

Family Isididae

Colonies in this family have a distinctly segmented axis that does not have sclerites incorporated in it (FIG. 9, p. 13). Although the calcareous internodes are usually solid, in some species they are hollow, but this tubular nature is not to be confused with the soft, cross-chambered central core of the holaxonians. Internodes can be coloured, and although they may be quite smooth, they are often ornamented with prickles and ridges. The nodes that alternate with the internodes consist of pure gorgonin.

Isis: p. 248

Colonies can be whip-like, but are usually profusely branched, bushy or fan-like, but rarely net-like. Polyps can be retractile or non-retractile. If non-retractile, they are commonly covered with broad scales, narrow rods or needles. If retractile, the polyps either have no sclerites or are armed with spindles in a collaret and points arrangement. Sclerites in the rest of the colony can be of many forms, including thorny or tuberculate spindles (often developed more on one side), 6- and 8-radiates, clubs, scales, and rooted-heads (which are rounded blocks with root-like structures on the bottom). The majority of species in this family are deep-water inhabitants.

Jasminisis: p. 250

Pteronisis: p. 252

Zignisis: p. 254

Photo: Roger Steene

Clavularia

Blainville, 1830

Colony shape: Colonies consist of individual, very large calyces, which are only connected at the base by stolons or thin membranes. In some cases, connecting stolons can be found growing from higher up the calicular wall. This may a species-specific character, or simply a micro-habitat adaptation.

Polyps: Monomorphic and retractile. In the ocean regions covered by this book they are commonly very large, with long, soft pinnules that are closely arranged and give the tentacles (which can be up to 20 mm in length) a distinct feathery appearance.

Sclerites: The pinnules and distal parts of the tentacles contain small, flattened discs or rods. The tentacle rachis contains smooth to warty rods. Although occasionally absent, the polyps usually have a strong points arrangement (sometimes a collaret) of long, narrow spindles, which are prickly or complexly warted, and sometimes branched. The calyx contains similar spindles that may be twice as long. The basal stolons or membranes contain large, warty and spiky, spindle-like sclerites, often branched and often fused in clumps. Rarely, similar colonies are found in which the sclerites of the calyx and stolons are small capstans or 6 - 8 radiates. Although in the literature such animals are included in this genus, further research is needed to establish the validity of this. Sclerites are always colourless.

Colour: Polyps pink or green to brownish-grey. Stolons brown to darkish-red, but often covered with turf algae or sponges. Zooxanthellate.

Habitat and abundance: On the Great Barrier Reef, *Clavularia* is common on near-shore reefs where it can cover extensive fields (hundreds of square meters). It can commonly be found on subtidal reef flats, and in other wave-exposed zones. Uncommon, and found in small clumps (never in extensive fields) in clear-water regions.

Zoogeographic distribution: Red Sea, South Africa, Korea, Palau, Guam, New Guinea, Great Barrier Reef, and other parts of the Indo-Pacific.

Similar Indo-Pacific genera: The expanded polyps could be mistaken for species of *Cervera*, *Stereosoma*, *Sansibia*, *Anthelia*, or *Tubipora*.

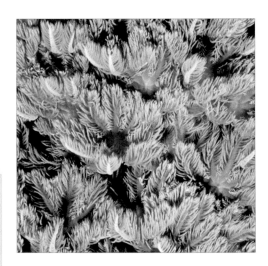

Points

Calyx

0.4 mm

Tentacle   Stolon   Fused clump from stolon

Photos:

ABOVE: In dense mats, stolons may be invisible when the polyps are expanded. Great Barrier Reef (GBR), ca 5 cm. *Photo: KF*

1: Fast-growing *Clavularia* can sometimes overgrow live coral. Central GBR, view ca 15 cm. *Photo: KF*

2: Extensive carpets of *Clavularia* (right) co-occur with *Briareum* (left) on coastal reefs of the GBR. *Photo: KF*

3 AND 4: Individual polyps, Papua New Guinea, view ca 5 cm. *Photos: Roger Steene*

5: Sabah, Malaysia, ca 10 cm. *Photo: Frances Dipper*

Cervera

López-González, Ocaña, Garcia-Gómez & Núñez, 1995

Colony shape: Small, plump polyps, growing close together on short, flattened stolons. Colonies are known to occur on coral rock or algae.

Polyps: Less than 10 mm tall when retracted. The upper part of the polyp can completely retract into the lower portion, which then becomes distended.

Sclerites: None

Colour: Brown. Most appear to be zooxanthellate.

Habitat and abundance: Only recently discovered in the tropics, but its apparent rarity may simply reflect the diminutive size of the colonies. Species from Japan are reported growing on the underside of rocks in shallow water.

Zoogeographic distribution: Known from the tropics from recent collections in Malaysia, Komodo, and the Solomon Islands. It also occurs in Japan.

Similar Indo-Pacific genera: *Clavularia.*

Remarks: This genus was first established for a shallow water, intertidal, species that occurs in the eastern Atlantic and throughout the Mediterranean. Included in the genus were some previously known species collected in Japan, at a latitude of about 35° North, which were originally described as species of *Cornularia.* Such a distribution casts some doubt on the identity of the tropical Indo-Pacific material. This doubt is further strengthened by the fact that both the Japanese and the European species would seem to be azooxanthellate, commonly growing underneath rocks. Unfortunately, these relatively simple animals have few features that can be used for identification, and the tropical Indo-Pacific material agrees fairly well with the general descriptions of the external characteristics of the cold-water species. We therefore place the central Indo-Pacific material in this genus only tentatively, until fresh specimens ideally prepared for histological examination allow further assessment. It is quite possible these warm water forms will prove to be species of *Clavularia* that do not have sclerites.

Photos:

ABOVE: *Cervera* from the Solomon Islands in a coral reef tank, view ca 15 cm. *Photo: Julian Sprung*

LEFT: Polyps in various stages of retraction, in what appears to be a *Cervera* (no sclerites are visible in the polyps). Komodo Island, Indonesia, view ca 5 cm. *Photo: Roger Steene*

Stereosoma

Hickson, 1894

Colony morphology: Stiff, well spaced, non-retractile polyps, united basally by thick, mat-like stolon.

Polyps: Monomorphic, non-retractile, non-contractile, and thick-walled. The polyp mouth is slit-like and situated on top of a small dome. The tentacles have a single row of 5 to 10 well-spaced pinnules along each side.

Sclerites: None.

Colour: Pale brown. Zooxanthellate.

Habitat and abundance: Described only as occurring along "shore reefs". Apparently rare, known only from one specimen.

Zoogeographic distribution: Southern part of Talisse Island, north Sulawesi.

Similar Indo-Pacific genera: *Clavularia, Cervera, Anthelia, Sansibia.*

Remarks: The validity of this genus remains a little uncertain because *Stereosoma celebense*, the only known species, has been recorded just once, in 1894, and, unfortunately, only a few fragments of the original specimen remain. Some authors suggested that the specimen was probably just a species of *Anthelia*, which is in the family Xeniidae, but *Stereosoma* has a full complement of eight mesenterial filaments, and so it is definitely not a xeniid. We have seen several published photographs purporting to be of *Stereosoma*, but the colonies are usually white, which indicates they have sclerites, and are probably species of *Anthelia* that have been misidentified.

The original description of *Stereosoma* reported that the polyps had a vacuolated, horny layer between the epidermis and the mesogloea, so the genus has been tentatively placed here, in the Family Clavulariidae.

Photo:

BELOW: Plate 45 *in*: Hickson, S.J. 1894. A revision of the genera of the Alcyonaria. Stolonifera, with a description of one new genus and several new species. Trans. Zool. Soc. Lond. 13 (9): 325-347

Carijoa
F. Müller, 1867

Colony shape: Tall, thin, axial polyps, which bud off lateral polyps and are united basally by a network of anastomosing stolons. A lateral polyp can sometimes grow like an axial polyp and bud off other lateral polyps, thus producing branched colony sections. Colonies can be up to 20 cm tall, and in most cases are overgrown by encrusting species of sponge or ascidian.

Polyps: Monomorphic, retractile, with a short body.

Sclerites: Slender and rod-like, ornamented with thorns and prickles, often branching and interlocking, and sometimes fusing into clumps. Colourless.

Colour: Polyps and stolons are white or cream, but encrusting sponges commonly cause colonies to appear in alternative colours, such as red, orange, yellow, blue or brown.

Habitat and abundance: Found predominantly in turbid coastal areas, most commonly as fouling organisms on jetties and wrecks. Also reported from shaded clearer waters in areas of strong currents.

Zoogeographic distribution: Widespread in both tropical and temperate waters.

Similar Indo-Pacific genera: *Paratelesto* and *Coelogorgia*.

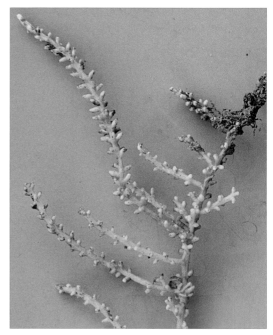

0.2 mm

Photos:

Top: A red sponge covers the branches of this *Carijoa*. Sabah, Malaysia, view ca 5 cm. *Photo: Frances Dipper*

Above: Dried colony from the GBR, ca 12 cm. *Photo: KF*

1: Sabah, Malaysia. *Photo: Frances Dipper*

2: Northern Great Barrier Reef (GBR), ca 10 cm. *Photo: KF*

3: Yongala Wreck, GBR, view ca 15 cm. *Photo: KF*

4: Papua New Guinea, view ca 10 cm. *Photo: Roger Steene*

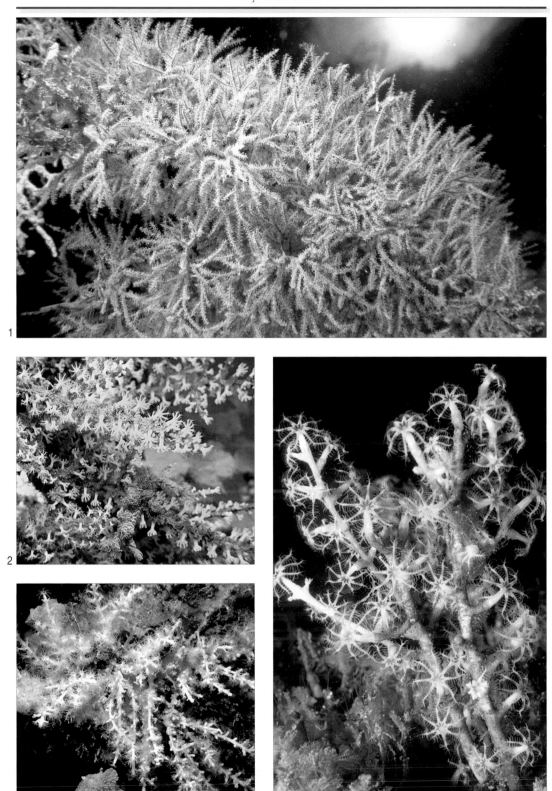

Paratelesto

Utinomi, 1958

Colony shape: The only shallow water Indo-Pacific species we have examined in detail grows as bushy clusters of somewhat stiff, laterally branched units that look like gorgonians. The stems are the thickened walls of axial polyps and bear small daughter polyps. Other long axial polyps, bearing small daughter polyps, branch from the stems up to at least the fourth order. The branched units are known to grow up to 16 cm tall, and they are united basally by a network of stolons overgrowing the substrate. A section through a thick branch shows small canal cavities arranged in several concentric rings around the larger gastric canal of the axial polyp.

Polyps: The polyps are retractile into narrow calyces, up to 1.5 mm tall, that are well spaced and arranged all around the stem and branches.

Sclerites: In the polyps, small, flattened spindles occur in a collaret and points arrangement, and the tentacles contain small, prickly rods. On the calyces there are oval spindles like those of the branch surface, and they are arranged in 8 longitudinal rows. In a branch several millimeters thick, the sclerites are arranged in 4 concentric layers. The surface or first layer contains large, complexly warted, oval spindles above a second layer of smaller, irregular shaped branched forms which

may fuse in clumps. Below this is a third layer of oval sclerites similar to those in the surface, and below this is a fourth layer of irregularly branched forms and fused clumps. Thicker branches have more layers. In the stem the many layers become confused. The sclerites are red.

Colour: Red, but the colour may be modified by a thin sponge overgrowth. Azooxanthellate.

Habitat: In the Semporna Islands, *Paratelesto* grows in a few current-exposed off-shore areas below 15 m depth.

Zoogeographic distribution: The only records are from the Semporna Islands in Sabah, Malaysia, from Bali in Indonesia, and Madang in Papua New Guinea, where apparently it is not uncommon. This would seem to indicate a distribution throughout the Indonesian waters.

Similar Indo-Pacific genera: *Carijoa, Coelogorgia,* and possibly *Siphonogorgia.*

Remarks: The only other known species occur in about 150 m of water off Japan. One is red, and the second is pink or pink with white calyces. In these species the sclerites are not reported to occur in concentric layers, though the canals are.

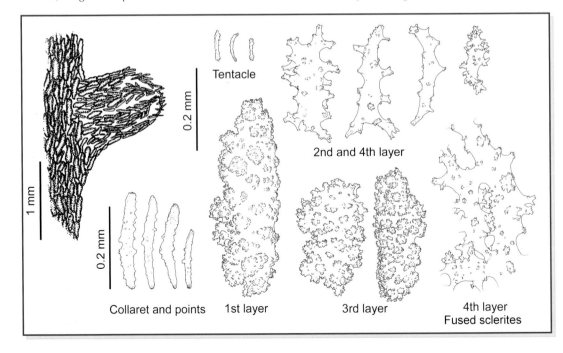

0.2 mm

1 mm

0.2 mm

Tentacle

2nd and 4th layer

Collaret and points   1st layer   3rd layer   4th layer
Fused sclerites

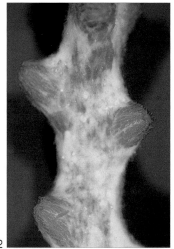

Photos:

1: Preserved specimens from Sabah, Malaysia, 14 cm tall. *Photo: PA*

2: Close-up of a preserved branch, showing sponge overgrowth, ca 0.5 cm. *Photo: PA*

3 AND 4: Sabah, Malaysia, ca 10 cm. *Photos: Frances Dipper*

5: Bali, Indonesia, ca 8 cm. *Photo: KF*

Coelogorgia
Milne Edwards & Haime, 1857

Colony shape: Colonies often appear as bushy clusters up to 1 m tall, and can be confused with gorgonians. Individual branched units are connected basally by a network of stolons overgrowing the substrate. Branches can arise in one plane or spirally. The main stems are actually the thick walls of the axial polyps. The axial polyps alternately bud off long, thin, daughter axial polyps that in turn bear short lateral polyps.

Polyps: Monomorphic, short, and non-retractile.

Sclerites: Small, lobed, granular scales in the polyp tentacles, and warty spindles throughout the rest of the colony. In the lateral branches and the upper parts of the stem the spindles are narrow and the warts usually in girdles. In the lower regions of the stems the spindles are thicker and the warts larger and often irregularly placed. Colourless.

Colour: Generally cream to white. Zooxanthellate.

Habitat and abundance: Little data is available. Probably restricted to sheltered waters, and uncommon.

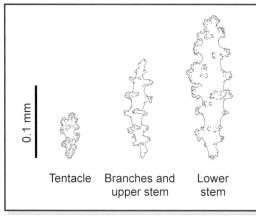

Tentacle   Branches and   Lower
            upper stem   stem

0.1 mm

Zoogeographic distribution: Zanzibar, Madagascar, Indonesia, Malaysia, and the Philippines.

Similar Indo-Pacific genera: *Carijoa* and *Paratelesto*.

Photos:

ABOVE: Large bushy *Coelogorgia* in the Philippines. *Photo: Gary Williams*

1: Philippines. *Photo: Coral Reef Research Foundation*

2: Colony growing in a sea water aquarium, ca 10 cm. *Photo: Julian Sprung*

3: Sabah, Malaysia, ca 8 cm. *Photo: Frances Dipper*

Tubipora

Linnaeus, 1758

Colony shape: Colonies are hemispherical and massive, to thick and encrusting, and consist of vertical, red, hard calcareous tubes ("organ pipes", hence its common name "organ pipe coral"), connected at various levels by horizontal, stolonic platforms. Colonies can be up to 50 cm in diameter, and each tube is formed and occupied by a single polyp, which is connected to the other polyps in the colony by canals inside the horizontal plates. When the colonies are alive, the skeleton is usually hidden by the tentacles of the extended polyps, and the gaps between tubes are often filled by sponges.

Polyps: Monomorphic and retractile, with relatively long bodies. The tentacles are arranged around a broad oral disc, and can vary from having a narrow rachis and long pinnules to a broad rachis and extremely reduced pinnules.

Sclerites: The tentacles contain minute corpuscle-like sclerites. The base of the anthocodia contains small, irregular, prickly rods, which may be branched or fused into small branching clumps. Similar sclerites are fused together to form the 'organ pipe' tubes of the colony.

Colour: Polyps cream-brown to grey-green to yellow-green to blueish-white, the oral disk sometimes being of a contrasting colour. The skeletal tubes are pale to dark red. Zooxanthellate.

Habitat and abundance: Common in a wide range of environments. On the Great Barrier Reef, it is most abundant on mid-shelf reefs, with abundances increasing with depth to about 20 m. Much less common in muddy coastal environments.

Zoogeographic distribution: Red Sea, Madagascar, Mozambique, most of the Indian Ocean and western Pacific (eastern limit: Marshall Islands and Vanuatu), and the Great Barrier Reef.

Similar Indo-Pacific genera: The expanded polyps may be confused with *Clavularia, Sansibia* and *Anthelia*.

Remarks: There are a number of nominal species in the literature, *T. musica* being the most well-known one. However, all are described from dried skeletons which vary in dimensions only, and need to be linked to specific polyp structures.

Red fragments can be a common sight on some coral beaches. However, regionally this genus is over-exploited, as the clusters of tubes are used for jewellery and ornaments.

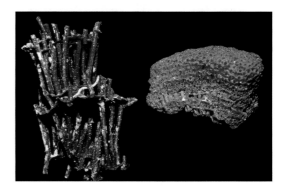

Photos:

TOP: Colony with unusually large polyps. Philippines, ca 8 cm. *Photo: Doug Fenner*

ABOVE: Dry skeletons of specimens with vastly varying tube sizes. Left: Philippines (the live colony is depicted in the TOP picture), right: southern Great Barrier Reef. *Photo: KF*

1: The "organ-pipe" structure of a *Tubipora* skeleton, here encrusted with coralline algae. GBR, ca 12 cm. *Photo: KF*

2: Whole colony. GBR, ca 12 cm. *Photo: KF*

3: Retracting polyps. GBR, ca 1.5 cm. *Photo: KF*

4: When polyps retract, the initial stages of newly formed horizontal plates become visible. Red Sea, Saudi Arabia, ca 2 cm. *Photo: Lyndon Devantier*

5: Two colonies with varying polyp sizes. Philippines, ca 8 cm. *Photo: Doug Fenner*

6 - 8: Varying tentacle shapes and pinnule lengths in *Tubipora*. GBR, ca 1.5 cm. *Photos: KF*

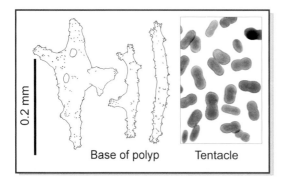

0.2 mm

Base of polyp   Tentacle

Sinularia

May, 1898

Colony shape: The variation in colony form seen in this genus is greater than in any other soft coral genus. Colonies may be low and encrusting with small knobs or ridges, tall and abundantly lobed, branched, flat and leaf-like or dish-like, and almost anything in between. In some species, a single, mature colony can cover tens of square meters, while other species are only several centimeters across. In turbid near-shore water, asexual reproduction by colony fission appears common, and aggregations consisting of hundreds of colonies can be found. Tightly contracted colonies are generally tough and hard, due to a dense mass of large, spindle-shaped sclerites in the colony interior (occasionally absent from the lobes). In some species, sclerites are fused together into a strong, rock-like substance called spiculite. Whole reef structures may be made from spiculite, and individual colonies of unknown age have been found sitting upon tall spiculite columns (**Photo 2**).

Polyps: Monomorphic, retractile, small, with short bodies. Tentacles are short, and arranged in a disk-shaped arrangement when expanded.

Sclerites: The surface of the polyp-bearing part of a colony characteristically contains well-formed clubs, sometimes long and thin, along with some small, narrow spindles or rods. There are only a few species in which there are no sclerites in this region. The surface of the stalk always contains clubs, and they are generally more robust versions of those found on the lobes. The interior of the lobes nearly always contains very large, complexly warted spindles (rarely ovals). There are only a few species where there are no sclerites in this region. The interior of the stalk always contains large complexly warted spindles (or, rarely, ovals) easily seen with the naked eye. The spindles in the interior regions of a colony are occasionally branched, sometimes to a remarkable degree. Sclerites are always colourless.

Colour: Brown, yellow, green, or cream. Polyps are usually the same colour as the general colony, but may be white or yellow. Zooxanthellate.

Habitat and abundance: Very abundant and widely distributed over most habitat types, from shallow waters down to at least 40 m, from very turbid to clean-water environments, and from very warm shallow bays into higher, cooler latitudes where reef growth ceases. Highly persistent species may dominate wave-protected coastal areas. Some species are tolerant of low light (steep walls, deep areas); others live on reef flats and tolerate intense illumination, storm waves, and exposure to air during extremely low tides.

Zoogeographic distribution: Widespread, from Africa and the Red Sea in the west to Hawaii in the east.

Similar Indo-Pacific genera: Some colonies may resemble *Cladiella*, *Lemnalia* or *Klyxum*. Confusion with species of *Lobophytum* is possible if no dimorphic polyps are visible (eg. on video), although the lobes in that genus are more regular in shape. *Dampia* has identical sclerites to *Sinularia* (see the remarks under that genus).

Remarks: In most species, the interior sclerites of *Sinularia* are over 2 mm long, and in a torn section they are easily seen with the naked eye. The outer surface of the lower base of live colonies feels rough due to these sclerites, distinguishing it from *Cladiella* and *Klyxum*.

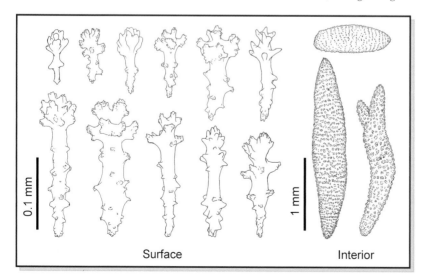

Surface   Interior

Photos:

1: Rich *Sinularia* assemblage on an island fringing reef of the central Great Barrier Reef (GBR). *Photo: KF*

2: *Sinularia* growing a solid calcareous framework. GBR, ca 1.5 m. *Photo: KF*

3: A branching species from deeper coastal waters of the GBR. Colony sizes ca 8 cm. *Photo: KF*

4: Large and slow-growing colonies, such as this one (measuring ca 8 x 10 m) from an off-shore reef of the northern GBR, may be hundreds of years old. *Photo: KF*

5

6

7

8

Photos:

5: *Sinularia* growing on an intertidal reef flat of Zanzibar. *Photo: Matt Richmond*

6, 9 AND 14: *Sinularia* (*cf S. polydactyla*) on near-shore island-fringing reefs of the GBR. *Photos: Carolina Bastidas*

7 AND 11: Off-shore species from the GBR. *Photos: KF*

8: *Sinularia brassica*, Palau. *Photo: Coral Reef Research Foundation*

10: A *Sinularia* carpet on a coastal reef of the GBR. *Photo: KF*

12: An arborescent species from Sabah, Malaysia. *Photo: Frances Dipper*

13: *Sinularia brassica*, with conspicuous surface sclerites and sparse polyps, is found in turbid deeper areas of the GBR. Ca 15 cm. *Photo: KF*

15: *Sinularia lamellata*, Palau. *Photo: Coral Reef Research Foundation*

Dampia

Alderslade, 1983

Colony shape: Large, low, encrusting colonies with large ridges on the upper surface. The surface always appears to be spiky because the ridges are covered with long calyces. The ridges, especially in the larger, expanded colonies, are usually parallel to each other, and more or less radially arranged. The ridge edge is irregularly digitate, and the ridges are often divided into sections that resemble a cock's comb. When the colonies are contracted, the ridges become sinuous, closely crowded together, and the tips of the finger-like lobes give the colony a lumpy appearance.

Polyps: Monomorphic and completely retractile into tall, narrow calyces. The calyces are crowded together on the lobed ridges, and very sparse on the surface between the ridges.

Sclerites: The colony surface contains a thin layer of club-shaped sclerites. The club head has a terminal wart with a whorl of 3 large warts below. The colony interior is densely packed with large, complexly warted spindles. The calyces are formed from the same type of spindles grouped together to form long tubes. Sclerites are colourless.

Colour: Yellowish or greenish-brown.

Habitat and abundance: Clear or turbid, sheltered, coral reef waters. Locally common.

Zoogeographic distribution: Reported from the Maldives, Thailand, the Philippines, Ambon, Sulawesi, the Rowley shoals and New Year Island off northern Australia, and the Great Barrier Reef.

Similar Indo-Pacific genera: The colony shape most closely resembles some species of *Lobophytum*, especially if the *Lobophytum* polyps are expanded to look like the calyces of *Dampia*.

Remarks: The sclerites of *Dampia pocilloporae-formis,* the only described species, are the same as those found in *Sinularia*, and it is possible that *Dampia* is simply a very aberrant species of that genus. Some colonies of *Sinularia brassica* are found with calyx-like protrusions, but not every polyp has them, and they are generally formed from thickened surface tissue, or from a few spindles laying on their sides and slightly proud of the surface. No *Sinularia* species are recorded in which the polyps have erect, discretely constructed, tubular calyces. Biochemical or molecular biological studies may help to resolve this uncertainty.

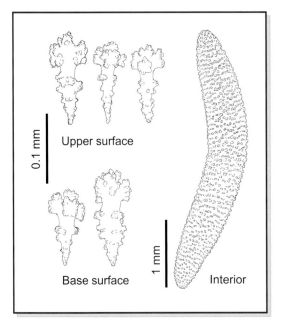

0.1 mm

Upper surface

Base surface

1 mm

Interior

Photos:
ABOVE AND 2: New Year Island, northern Australia. *Photos: PA*
1: Northern GBR, ca 15 cm. *Photo: KF*

Cladiella
Gray, 1869

Colony shape: Encrusting to erect, firm to fleshy, and lobate. Colonies are commonly relatively small and have limited powers of contraction. Asexual reproduction is by colony fission.

Polyps: Monomorphic, retractile, small, with short bodies. When expanded, the tentacle tips bend far back to form an inverse bowl shape.

Sclerites: All very small. Those of the polyps are generally minute disks or figure-eights, and roughended dumbbells or double-heads. Those of the rest of the colony are dumbbells with rounded or cone-shaped processes. As the sclerites are so small, the colony surface generally feels smooth and soft. Sclerites are colourless.

Colour: Colonies are generally white, grey or pale brown when contracted. Polyps are brown or greenish due to dense concentrations of zooxanthellae. The rapid retraction of polyps upon disturbance results in a characteristic instantaneous colour change from brown to white.

Habitat and abundance: Moderately common in coastal waters on crests and rocky coasts in wave-exposed habitats. Uncommon in a wide range of other reef habitats. They are found as individuals or clones of a small number of evenly sized colonies.

Zoogeographic distribution: Widespread, from Africa and the Red Sea in the west, to the western Pacific islands in the east.

Similar Indo-Pacific genera: Some colonies may resemble *Sinularia*. However, the lobes of *Cladiella* are generally shorter and more densely packed, and never form ridges, and the surface of the base of *Cladiella* never feels rough. Other species may be mistaken for *Klyxum* and require an examination of the sclerites for differentiation.

Photos:

BELOW: Central GBR, ca 10 cm. *Photo: KF*

1: Clusters of *Cladiella* are commonly found in wave-exposed shallow water on rocky substrate. Great Barrier Reef (GBR), view ca 25 cm. *Photo: KF*

2: Swain Reefs, Southern GBR, ca 15 cm. *Photo: PA*

3: Intertidal colony from Gladstone, southern GBR, ca 12 cm. *Photo: PA*

4: Hong Kong, ca 15 cm. *Photo: KF*

5: *Cladiella* grazed upon by an egg cowry (*Ovula ovum*). Kenia, ca 15 cm. *Photo: Mark Wunsch*

Polyp

0.1 mm

Around polyp opening

Surface and interior

Klyxum

Alderslade, 2000

Colony shape: Lobate colonies, nearly always soft and fleshy, with long, partly subdivided lobes which are evenly covered with polyps. Most species are small, but several species can grow quite large (greater than 1 m). Colonies are very soft, highly contractile and may appear slightly translucent, because the number of sclerites in the branches is generally very small and the water content high. Asexual reproduction by colony fission appears to be uncommon.

Polyps: Monomorphic, small, with short bodies. They appear to be retractile but they are actually highly contractile, with the ability to deflate until flush with the lobe surface.

Sclerites: The characteristic sclerites are spindles that have large cone-shaped prominences. In many species these are only found in the interior of the colony base. In others they are also found in the interior of the lobes. Sometimes the interior of the lobes only contains a few very thin, granular rods. Polyp sclerites are minute granular disks and flattened rods, which may be rare or very numerous. Colourless.

Colour: The polyps are usually medium to dark brown, while the colour of contracted colonies is lighter (cream, or pinkish brown). In some species both polyps and colony are brown. Zooxanthellate.

Habitat and abundance: Moderately common in turbid coastal waters protected from wave-exposure. Infrequent in other habitats of the Great Barrier Reef. Mostly found as individual colonies, occasionally in small groups of a few colonies.

Zoogeographic distribution: Recorded from Madagascar, the Red Sea, Thailand, Indonesia, Philippines, Japan, Palau, Papua New Guinea, the Great Barrier Reef, New Caledonia and Fiji.

Similar Indo-Pacific genera: Some colonies closely resemble *Cladiella*, however the lobes are generally more translucent, fleshy, and longer. The larger species could be mistaken for *Sinularia*. Some species resemble *Efflatounaria*.

Remarks: Tropical Indo-Pacific species of this genus have previously been classified as *Alcyonium* Linnaeus, 1758. However recent research has shown that this classification is incorrect.

Photos:

BELOW: Close-up of a preserved colony, translucent due to the sparsity of sclerites in the colony surface, with clearly visible sclerite rods in the polyps. Great Barrier Reef (GBR), view ca 1.5 cm. *Photo: KF*

1: Unusually large (> 1 m diameter) colony from a near-shore reef of the central GBR. *Photo: KF*

2, 6, AND 8: GBR, views ca 10 cm. *Photos: KF*

3: Saudi Arabia, Red Sea. View ca 10 cm. *Photo: Lyndon Devantier*

4: Northern GBR, view ca 5 cm. *Photo: KF*

7: Southern GBR, ca 15 cm. *Photo: PA*

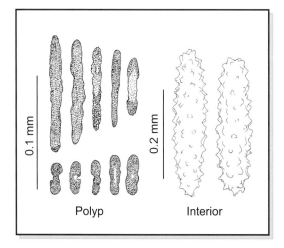

Polyp   Interior

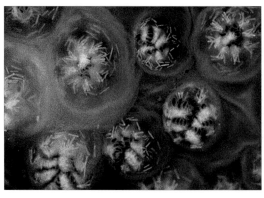

Rhytisma

Alderslade, 2000

Colony shape: Thin-encrusting mats or ribbons of irregular shape, about 2 - 4 mm thick, often overgrowing dead, sometimes live, substrate. The large gastric cavities of the polyps extend nearly right through the mat or basal membrane. Asexual reproduction is by colony fragmentation.

Polyps: Monomorphic, retractile, and of medium size with a short body.

Sclerites: The colonies contain large spindles, up to several millimeters long, sometimes appearing as a conspicuous honeycomb-like network on the upper surface with one polyp within each 'cell'. The polyps also contain spindles arranged in a collaret and points formation. Colourless.

Colour: Colour is variable. The encrusting mat can be greyish-purple, greenish-yellow, cream or pale brown. The polyps are greyish-purple, intense greenish-yellow, cream, orange, or brown. Zooxanthellate.

Habitat and abundance: Uncommon, but patchy (can be abundant in a large area). On the Great Barrier Reef, it is most frequently found at 5 - 15 m depths on mid-shelf reefs, but occurs all across the shelf except on very turbid coastal reefs.

Zoogeographic distribution: Widespread and likely to be found in most areas covered by this book.

Similar Indo-Pacific genera: Some pale purple forms may superficially resemble *Briareum*.

Remarks: Species of this genus have previously been classified as *Parerythropodium* Kükenthal, 1916. However, recent research has shown that this classification is incorrect.

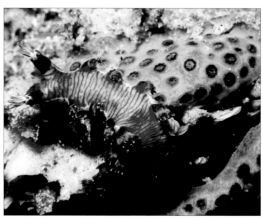

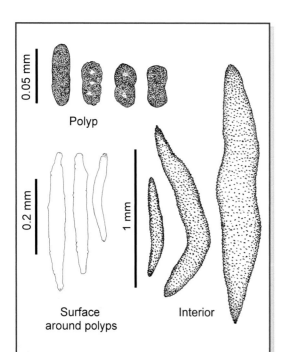

Photos:

Above: *Rhytisma*, preyed upon by a nudibranch. Red Sea, ca 5 cm. *Photo: Mark Wunsch*

1: Central Great Barrier Reef (GBR), ca 50 cm. *Photo: KF*

2: Zanzibar, ca 15 cm. *Photo: Matt Richmond*

3: Red Sea, ca 15 cm. *Photo: Mark Wunsch*

4: Central GBR, ca 15 cm. *Photo: KF*

5: Northern GBR, ca 80 cm. *Photo: KF*

6: Close-up of honeycomb-like spindle formation around polyps of a preserved colony. View ca 1 cm. *Photo: PA*

7: Half-contracted polyps, central GBR, ca 4 cm. *Photo: KF*

8: A contracted colony with clearly visible surface spindles. Central GBR, ca 4 cm. *Photo: KF*

Sarcophyton

Lesson, 1834

Colony shape: Colonies have a conspicuous bare stalk that merges into a wider, fleshy, disk-like polyp-bearing region called polypary. Polyps are found only on the upper side of the polypary. The polypary of adults is concave in the centre, and folded or wavy around the periphery. Juveniles have a convex, non-folded polypary and look like mushrooms. Colonies are small to over 1.5 m in diameter, soft and fleshy, with extensive powers of contraction. Stalks grow short and thick in wave-exposed environments, and long and thin in deep mud and soft-bottom habitats. Asexual reproduction by colony fission, or the budding of daughter colonies off the stalk or the edge of the polypary, is common in turbid water, but rare in a clear-water environment. Colonies of some species (eg, *S. tortuosum*) are able to migrate over distances of ~ 0.5 m within a few months.

Polyps: Dimorphic. Expanded autozooids have very long bodies (up to 20 mm), with a medium-sized oral disk and tentacles. They are completely retractile, leaving the colony surface looking smooth and leathery. The siphonozooids are small, very numerous, and densely arranged between the autozooids.

Sclerites: Sclerites of the surface of the polypary and stalk are characteristically well-formed clubs, occasionally very long and thin. The colony interior contains sticks and spindles. Those in the stalk are usually thicker and more robust than those in the polypary. Colourless.

Colour: Brown, beige, yellow, green, or cream. The autozooid polyps are usually the same colour as the general colony, but may be yellow or white in some brown colonies. Zooxanthellate.

Habitat and abundance: Very abundant and widely distributed, from the intertidal to considerable depths, and from very muddy coastal to off-shore environments. On moderately turbid near-shore reefs and soft bottom habitats, it is not uncommon to find extensive clones consisting of many hundreds of colonies, which appear to display fast rates of growth and asexual reproduction. In contrast, colonies tend to be slow-growing on clear-water reefs.

Zoogeographic distribution: Widespread, from eastern Africa and the Red Sea in the west, to Polynesia in the east.

Similar Indo-Pacific genera: In wave-exposed habitats, flat colonies with very short and thick stalks may resemble several species of *Lobophytum*, but the folds of the disk are rarely modified as lobes. It should be noted however, that atypical species of *Sarcophyton* look just like atypical species of *Lobophytum*, and sclerite examination is required for differentiation.

0.1 mm

Surface   Stalk   Polypary
           interior   interior

Photos:

1: The typical shape of *Sarcophyton*. Solomon Islands. *Photo: Roger Steene*

2: Colonies of this species (*S. tortuosum*) can migrate by directional growth. Central Great Barrier Reef (GBR), 50 cm. *Photo: KF*

3: Almost contracted autozooids. GBR, ca 4 cm. *Photo: KF*

4: Expanded autozooids. The siphonozooids are also visible, appearing as darker dots on the surface between the autozooids. GBR, ca 4 cm. *Photo: KF*

5 AND 6: GBR, ca 30 cm. *Photos: KF*

Photos:

7: Eaten colony with some initial signs of regrowth of the polypary. Great Barrier Reef (GBR), ca 10 cm. *Photo: KF*

8: Partly eaten colony, with first signs of polyp regeneration on the right edge of the polypary. GBR, ca 15 cm. *Photo: KF*

9: Colony with numerous commensals (Plathelmintha: probably *Convolutriloba retrogemma*). GBR, ca 15 cm. *Photo: KF*

10: Small colonies budding off the edge of the polypary. Central GBR on a coastal reef, ca 50 cm. *Photo: KF*

11: Colony inhabited by translucent-yellowish sedentary ctenophores (*Coeloplana* sp.), which use their extended tentacles for filter-feeding. GBR, ca 20 cm. *Photo: KF*

12 AND 13: Central GBR. *Photos: KF*

14: A coastal species, central GBR, ca 60 cm. *Photo: KF*

15: Sabah, Malaysia, view ca 150 cm. *Photo: Frances Dipper*

16: Intertidal colony, northern Australia. *Photo: PA*

17: Juvenile colony, Red Sea, ca 3 cm. *Photo: Mark Wunsch*

12

13

14

15

16

17

Lobophytum

Marenzeller, 1886

Colony shape: Colonies are thick-encrusting, firm to touch, with polyps present only on the upper surface. The upper surface is about the same diameter as the colony base, and generally has lobes and ridges. The lobes are nearly always simple and unbranched, but ridges can occur that look like cock's-combs. Ridges tend to be oriented parallel or radially in a relatively regular manner. In some coastal species, lobes are very slender and prominent thus colonies appear digitate. A few species are flat and free of lobes or ridges on the surface. Low, encrusting colonies can grow up to a square meter in area. Asexual reproduction is by colony fission. Colonies may have considerable powers of contraction.

Polyps: Dimorphic. Expanded autozooids have long bodies (up to 10 mm), with a medium-sized oral disc and tentacles. They are completely retractile, leaving the colony surface looking smooth and leathery. The siphonozooids are small, very numerous, and densely arranged between the autozooids. In off-shore species, polyps are rarely expanded during the day.

Sclerites: Sclerites of the surface of the polyp region and stalk are characteristically poorly-formed clubs. The interior of the lobes contains spindles. The sclerites of the interior of the base are usually thicker and more robust than those in the lobes, and are spindles or ovals with the warts nearly always arranged in several girdles. Although the interior

sclerites may be up to 0.5 mm long, they are more commonly about half that length, making the feel and appearance quite different from that of *Sinularia*. Colourless.

Colour: Commonly a characteristic yellow-brown, with the tips of lobes often lighter than the sides. A few species are greenish or cream-brown. Autozooid polyps are normally white, yellow, or greenish-brown. Zooxanthellate.

Habitat and abundance: Common in shallow, well-lit water on reef flats, crests and upper slopes, and down to 25 m in very clear water. Uncommon on steep slopes and in lagoons. Rare in muddy environments. On clear-water reefs, colonies generally occur as individuals. Some digitate near-shore species grow in aggregations consisting of tens of colonies.

Zoogeographic distribution: Widespread, from east Africa and the Red Sea, to Polynesia.

Similar Indo-Pacific genera: Some *Sinularia* species have undivided lobes, thus some confusion could occur if siphonozooids are invisible (eg. on low-resolution video and photographs). Some confusion could also occur with colonies of *Sarcophyton*, see remarks under that genus. Colonies with radially arranged lobes and half-expanded polyps can resemble *Dampia*, because the polyps can look like the calyces present in that genus.

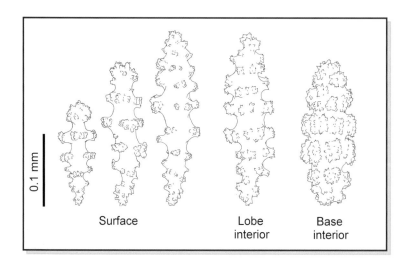

Surface   Lobe interior   Base interior

0.1 mm

Photos:

1: A close look clearly shows the two types of polyps in *Lobophytum*. Great Barrier Reef (GBR), ca 4 cm. *Photo: KF*

2: Fleshy lobes are commonly found in large coastal species on the GBR. View ca 15 cm. *Photo: KF*

3: Many alcyoniid corals shed mucus layers to clean the colony surface and remove sediments. GBR, ca 15 cm. *Photo: KF*

4: Several species of *Lobophytum* have flat surfaces (here: a *L. depressum* from Hong Kong, ca 20 cm). *Photo: KF*

5: An off-shore species from the GBR. Ca 12 cm. *Photo: KF*

6

7

8

9

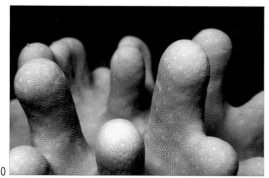

10

11

Photos:

6: *Lobophytum*, grazed upon by cowries of the genus *Ovula*. Kenia, ca 40 cm. *Photo: Mark Wunsch*

7: Great Barrier Reef (GBR), ca 40 cm. *Photo: KF*

8: Pits of retracted autozooids among numerous small siphonozooids. GBR, ca 3 cm. *Photo: KF*

9: GBR, ca 70 cm. *Photo: PA*

10: Contracted off-shore species, GBR, ca 5 cm. *Photo: KF*

11: A lobate coastal species overgrowing a massive *Porites*. GBR, ca 40 cm. *Photo: KF*

12 AND 15: Coastal *Lobophytum*. GBR, *Photos: KF*

13 AND 14: GBR, 15 - 20 cm. *Photos: KF*

16: Clear-water colony, GBR, ca 10 cm. *Photo: KF*

12

13

14

15

16

Paraminabea

Williams & Alderslade, 1999

Colony shape: Small, dome-shaped to carrot-shaped colonies, rarely sparsely branched. Colonies highly contractile and up to 10 cm long when expanded.

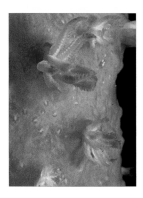

Polyps: Dimorphic. Autozooids are large, retractile, and have relatively long bodies. Siphonozooids are tiny, and scattered between the autozooids. It is commonly necessary to cut across the polypary or remove a thin slice from the surface in order to detect the presence of siphonozooids in preserved material.

Sclerites: Surface and interior sclerites are 6- or 8-radiates, spindles derived from these forms, and tuberculate spheroids. The polyps are free of sclerites. The sclerites are usually coloured.

Colour: Orange, red, yellow, or pinkish-white to cream-coloured. Polyps white or orange. Azooxanthellate.

Habitat and abundance: Uncommon, and restricted to darker areas such as caves and overhangs, often below 10 m depth. Generally found in small groups of a few colonies.

Zoogeographic distribution: Known from south east Africa, Madagascar, Seychelles, Maldives, Sri Lanka, Hong Kong, Philippines, Japan, Micronesia, Indonesia, New Guinea, Great Barrier Reef, Solomon Islands, and Fiji.

Similar Indo-Pacific genera: *Eleutherobia* and *Bellonella,* which have monomorphic polyps.

Remarks: All the shallow-water species previously assigned to the genus *Minabea* Utinomi, 1957 are in fact *Paraminabea. Minabea* only occurs in deep water off Japan and New Zealand.

Photos:

LEFT: Small, slit-shaped siphonozooid openings are visible at close inspection amongst the larger autozooids. Central Great Barrier Reef (GBR), ca 2 cm. *Photo: KF*

BELOW: The beautiful autozooids are mostly expanded at night. Central GBR, ca 3 cm. *Photo: KF*

1: Overhang and walls are the preferred habitat of *Paraminabea.* Central GBR. *Photo: KF*

2: Solomon Islands, ca 10 cm. *Photo: Gary Williams*

3: Ca 5 cm. *Photo: Roger Steene*

4 AND 5: A so far undescribed branched species from Hong Kong, ca 5 cm. *Photos: KF*

6: The recently described *Paraminabea arborea* from the Philippines. *Photo: Coral Reef Research Foundation*

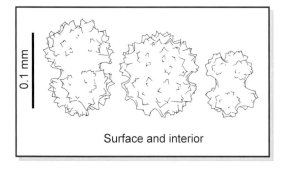

0.1 mm

Surface and interior

Eleutherobia

Pütter, 1900

Colony shape: Cylindrical, sausage- or carrot-shaped colonies, rarely branched, and up to about 9 cm long when contracted. The polyp-free basal portion is usually very short, but it can occupy up to half of the colony. Calyces, large to very small, are clustered on the upper, polyp-bearing portion of the colony.

Polyps: Monomorphic, large and retractile with very long bodies. Generally expanded at night and contracted during the day.

Sclerites: In a few species the polyps are devoid of sclerites, but in most, the polyp head contains numerous straight or curved spindles in a collaret and points arrangement. The tentacles contain small to large rods, often branched, and flattened spindles, usually curved. The polyp body contains many small spindles and clubs. The surface of the calyces, polypary, and base usually contain small 8-radiates and capstans (or modified forms of these), and rarely warty spheroids. The colony interior may contain similarly shaped sclerites, spindles, branched forms, or some combination. Sclerites are commonly coloured.

Colour: Very variable - pink and red, white, white with red calyces, red, grey with red or brown calyces, yellow, orange with yellow or red calyces, and brown. A species may occur in several colour forms. Azooxanthellate

Habitat and abundance: The few species known from shallow water are generally restricted to caves and overhangs where they are found in small groups of a few colonies. Most known species grow below 45 m, and have been collected by dredging, which indicates they may live on open sea bottom.

Zoogeographic distribution: Most records are from Japan, in waters beyond normal diving depths (the shallowest were 24 and 40 m). Deep-water records exist also from South Africa, Somalia, Western Australia, Indonesia and the Philippines. The few shallow water records originate from the Great Barrier Reef, Western Australia, Papua New Guinea, Palau, Korea, and the Philippines.

Similar Indo-Pacific genera: *Paraminabea*, which has dimorphic polyps.

Remarks: This genus contains nominal species that exhibit a wide variety of characters. Further work will clarify if some of these species would be better placed in other genera.

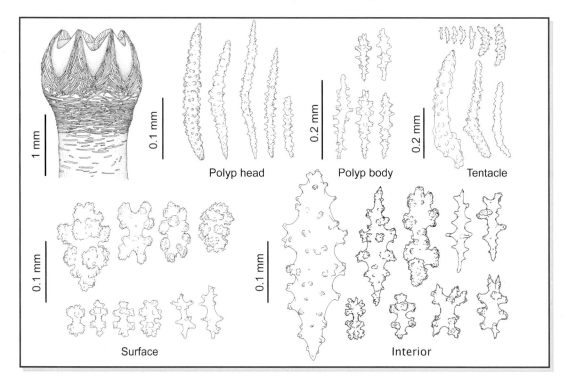

Polyp head

Polyp body

Tentacle

Surface

Interior

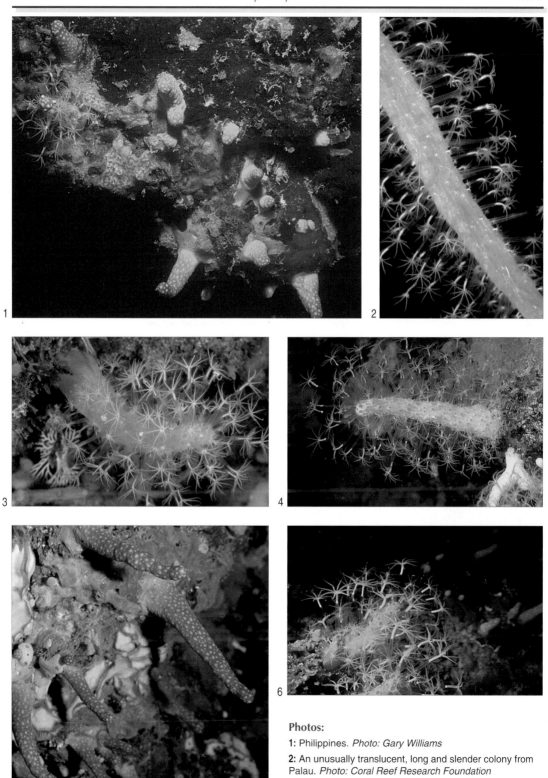

Photos:

1: Philippines. *Photo: Gary Williams*

2: An unusually translucent, long and slender colony from Palau. *Photo: Coral Reef Research Foundation*

3 - 5: Philippines. *Photos: Coral Reef Research Foundation*

6: Philippines. *Photo: Leen van Ofwegen*

Bellonella

Gray, 1862

Colony shape: This genus was established in 1862 for a single specimen, named *Bellonella granulata*, which was collected off northern Australia. It is the only species known from a diveable depth, and has never been collected again. The original colony, shown in Fig. 1, is 2.5 cm high, and torch-shaped. There are a number of large calyces on the summit, and some polyps that are extended.

A number of other species, all from deep water, have been placed in this genus, possibly erroneously. Most are small, finger–shaped colonies, rarely branched, and a couple are long and whip-like. The description below refers to *Bellonella granulata*.

Polyps: Monomorphic, large, and retractile into well-developed, thick-walled calyces.

Sclerites: In the surface of the colony and the walls of the calyces there is a thick layer of small sclerites, shaped like clubs, narrow spindles, or needles, with simple cone-like prominences. The interior of the colony contains longer, narrow spindles with a similar architecture. The polyp head contains numerous narrow spindles, like those found in the colony interior, which are densely placed in a collaret and points arrangement. Colourless.

Colour: Not recorded.

Habitat and abundance: Very rare. The type specimen was found in 31m of water.

Zoogeographic distribution: Bellonella Bank, Timor Sea.

Similar Indo-Pacific genera: *Nidalia*.

Remarks: The distribution, and the variety of colony form and sclerite shape found amongst the other nominal species suggest that many may be misplaced in this genus. Surface sclerites of the polyp zone in these other species are often just spindles, and many have a separate stalk, below the cylindrical polyp area, which has a different sclerite shape entirely. Indian and Pacific Ocean deep-water species are recorded from off Madagascar, Philippines, and Hawaii. The remaining species are from the Mediterranean and the Atlantic. In some species, all or some of the sclerites are red.

Photo:
1: The original, preserved specimen of *Bellonella granulata*, 2.5 cm tall. *Photo: PA*

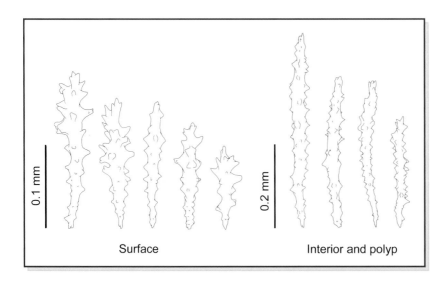

0.1 mm

0.2 mm

Surface

Interior and polyp

1

Nephthea
Audouin, 1826

Colony shape: Bush-like to tree-like, up to 0.5 m tall, and quite soft.

Polyps: Monomorphic, non-retractile, and clustered at the end of the terminal branches, forming catkins or lobes.

Sclerites: In the polyp head there are irregularly arranged bent spindles that have the outward-pointing end most developed. The longest sclerites occur on the outer side of the polyp, the smallest on the inner side. A supporting bundle is present, but commonly not large and conspicuous. Sclerites of the surface layer of the branches are spindles, often with spines along one side. The stalk surface has similar sclerites along with numerous capstans and derivations of capstans. Sparsely tuberculated spindles occur in the colony interior. Sclerites are always colourless.

Colour: Cream to brownish-yellow, brownish-green or brownish-purple. Zooxanthellate.

Habitat and abundance: Locally abundant, particularly in wave-protected clear-water areas. Rare in wave-exposed areas and absent in very turbid water.

Zoogeographic distribution: Widespread and could reasonably be expected to be found in most areas covered by this book.

Similar Indo-Pacific genera: *Litophyton, Stereonephthya* and *Lemnalia.*

Remarks: Many species of *Nephthea* with coloured sclerites have been described in the scientific literature. Current research has shown that some of these belong to *Stereonephthya*, while others belong to an as yet undescribed genus.

Photos:
BELOW: *Nephthea* growing on a live *Acropora*. Great Barrier Reef (GBR). *Photo: KF*
1 - 6: A variety of species from the GBR. *Photos: KF*
7: Palau. *Photo: KF*

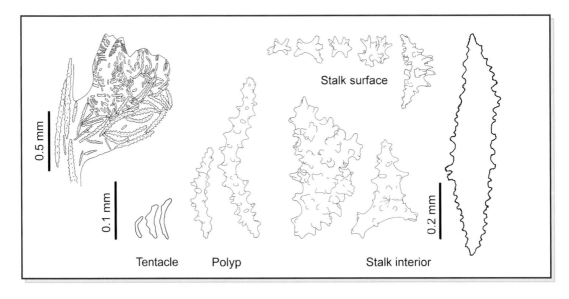

Tentacle Polyp   Stalk surface   Stalk interior

Litophyton

Forskål, 1775

Colony shape: Bush-like to tree-like, up to 0.5 m tall, and very soft.

Polyps: Monomorphic, non-retractile, and clustered at the end of the terminal branches, forming catkins.

Sclerites: Polyps with small, irregularly arranged rods, or completely unarmed. Sclerites of the surface layer of the branches are spindles, often with spines along one side. In the surface of the stalk similar spindles occur along with capstans and derivations of capstans. In the interior of the stalk there are sparsely tuberculated spindles. Always colourless.

Colour: Yellow, cream to brown, or purple. Zooxanthellate.

Habitat and abundance: Locally common in the Indian Ocean and the Red Sea, particularly in wave-protected areas. Rare on the GBR.

Zoogeographic distribution: Widespread and could reasonably be expected to be found in most areas covered by this book.

Similar Indo-Pacific genera: *Nephthea* and *Lemnalia*. Because of the sparse and small sclerites in the polyps and branches, colonies of *Litophyton* are generally far more flabby than those of *Nephthea* and *Stereonephthya*. There are some species where an apparent overlap of characters makes it very difficult to decide if they should be called *Nephthea* or *Litophyton*. Whether the two genera should be united is currently being examined.

Photos:
1: Sabah, Malaysia, ca 30 cm. *Photo: Frances Dipper*
2: Philippines. *Photo: Coral Reef Research Foundation*
3: Red Sea, ca 50 cm. *Photo: Mark Wunsch*
4: Great Barrier Reef, view ca 8 cm. *Photo: KF*
5: Palau. *Photo: Coral Reef Research Foundation*
6: A large cluster of colonies (about 50 cm tall) from the Great Barrier Reef. *Photo: KF*
7: Zanzibar. *Photo: Matt Richmond*

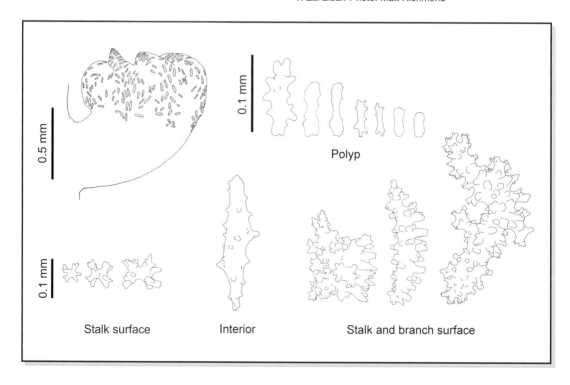

0.5 mm

0.1 mm

0.1 mm

Polyp

Stalk surface

Interior

Stalk and branch surface

Stereonephthya

Kükenthal, 1905

Colony shape: Bush-like to tree-like, often rough to the touch and relatively small (< 20 cm).

Polyps: Monomorphic and non-retractile, the head often making an acute angle with the polyp body. They arise singly or in small groups from the terminal branches, and sometimes also from the stalk and main branches. Catkins also occasionally occur.

Sclerites: Polyps each have a conspicuous supporting bundle. One or more of the spindles of the supporting bundle generally project beyond the polyp head. Sclerites on the polyp head are bent spindles with the outward pointing end more developed than the inner end. The largest spindles occur on the outward-facing side of the polyp, the smallest on the inner side. They can be organised as 8 points, but mostly they are irregularly arranged. Sometimes some of the dorsal spindles project beyond the polyp head. On the ventral side of the polyp body, small rods occur. Sometimes these rods are present over the complete polyp body, and may even continue onto the branch surface. Sclerites of the surface layer of the branches are spindles often with spines along one

side. The stalk surface has similar sclerites along with numerous capstans and derivations of capstans. Sparsely tuberculated spindles occur in the colony interior. Colourless or coloured.

Colour: White, red, yellow, and pink. Sometimes in one colony the stalk, the branches and the polyps each have different colours. Azooxanthellate.

Habitat and abundance: Occasionally encountered in both clear and turbid water.

Zoogeographic distribution: Widespread and could reasonably be expected to be found in most areas covered by this book.

Similar Indo-Pacific genera: *Dendronephthya, Nephthea,* and *Litophyton.*

Photos:

1: Great Barrier Reef (GBR), view ca 20 cm. *Photo: KF*

2: Palau, ca 10 cm. *Photo: KF*

3: GBR, colony size ca 12 cm. *Photo: KF*

4 AND 7: Palau, view ca 10 cm. *Photos: Coral Reef Research Foundation*

5 AND 6: Preserved colonies from the GBR, view ca 2 cm. *Photos: KF*

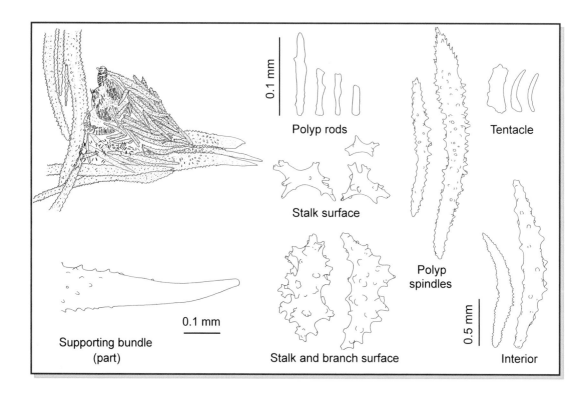

0.1 mm

Polyp rods

Tentacle

Stalk surface

Polyp spindles

0.1 mm

Supporting bundle (part)

Stalk and branch surface

0.5 mm

Interior

Scleronephthya

Studer, 1887

Colony shape: Colonies generally small (<10 cm), but at least one Indonesian species grows to > 0.5 m tall. Highly contractile, often appearing as coloured, lumpy crusts when contracted, but when expanded they are stout and arborescent, often planar, with sparsely subdivided branches or lobes.

Polyps: Monomorphic and small, highly contractile, and only found in the branched portion of the colony. Polyps occur as isolated individuals, rather than in groups or bunches. However, polyps can appear to be grouped on small lobes in contracted colonies. Usually expanded at night or in strong currents.

Sclerites: Conspicuous points arrangement in the polyp heads. Polyp body commonly with spindles of different sizes angled around it, which may come together to form a pseudo-collaret in contracted polyps. The points contain sticks and spindles of various sizes, some thickened club-like, or pointed, at one end. The sclerites of the pseudo-collaret and polyp body are often large and curved to fit around the polyp. Tentacles contain small, knobby rods. Colony surface and stalk interior contain large, warty spindles, occasionally branched. Most of the larger surface and interior sclerites have high, rounded warts covered in fine prickles, the shape of which is very characteristic of the genus. At least one species has a fine 'dusting' of minute, flattened platelets covering the polyp

region, and some Indonesian species have hardly any sclerites in the polyps. Sclerites are always colourless.

Colour: The most common colour is orange, usually with pale yellow, pink, or even green stalk and branches. However, some species are entirely pink, and one Maldives species is white with yellow polyps. Azooxanthellate.

Habitat and abundance: Found in flow-exposed areas, on steep walls, wrecks, and overhangs or cave entrances, commonly in groups of many tens of colonies. In Australia they are uncommon, but occur in the whole range from turbid inshore waters to outer clear-water reefs. Colonies have been dredged from 30 to 85 m depths. Locally abundant in the northern Red Sea.

Zoogeographic distribution: Red Sea, Korea, Taiwan, Japan, Philippines, Indonesia, Papua New Guinea, Australia, and the Solomon Islands.

Similar Indo-Pacific genera: *Dendronephthya, Nephthea, Stereonephthya.*

Photos:

1: Current-swept wall covered with *Scleronephthya*. Gulf of Aqaba, Red Sea. *Photo: KF*

2 AND 6: Northern Great Barrier Reef, 5 & 10 cm. *Photos: KF*

3: Maldives. *Photo: Coral Reef Research Foundation*

4: Northern Red Sea, ca 5 cm. *Photo: KF*

5: Palau. *Photo: Coral Reef Research Foundation*

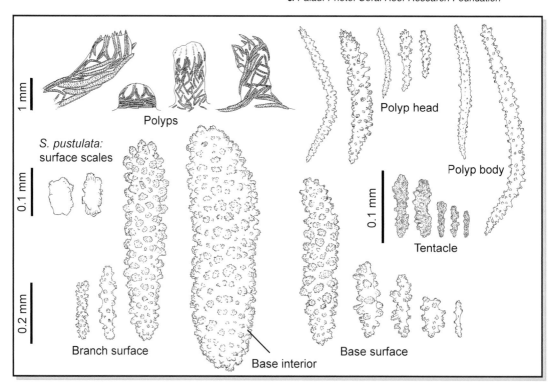

Polyps

S. pustulata: surface scales

Polyp head

Polyp body

Tentacle

Branch surface

Base interior

Base surface

1 mm

0.1 mm

0.2 mm

0.1 mm

Dendronephthya

Kükenthal, 1905

Colony shape: Highly branched or bushy colonies, where the end branches and polyp bunches are generally arranged in one of three growth forms (though intermediate forms are not uncommon) whose taxonomic value is unclear:

DIVARICATE: sparse, arborescent branching, with slender ramifications, and polyps in small insignificant, bundles (e.g., **PHOTO 2**).

GLOMERATE: close, short branching, where the polyp bundles are assembled to form large, distinctly rounded bunches (e.g., **PHOTOS 10 AND 3**).

UMBELLATE: polyp bundles closely arranged at the same level on the ends of the twigs, forming numerous umbrella-like crowns (e.g., **PHOTO 6**). Several umbrella-like sections may combine to form hemispheres.

Polyps characteristically grow in small bundles of about 2 - 15 polyps predominantly on the terminal twigs (**PHOTO 4**), but occasionally on the larger branches (**PHOTO 12**). Very large, white or coloured spindle-shaped sclerites can be seen in the branch and twig surfaces. The length of the basal, bare stalk is quite variable, and a collar-like 'branch', which usually shelters a pair of shrimps, commonly occurs at the top of the stalk forming the first part of the polyp-bearing zone. Colonies are usually around 20 cm tall, but some as large as 2 m have been found. Colonies are generally expanded at night, and in the day during periods of high current. When contracted, they can shrink to just a small proportion of their extended size. Some species form rhizoids (root-like extensions of the stalk) for colony attachment. *(continued)* ➡

Photos:

1: Current-swept wall with multi-coloured *Dendronephthya* and yellow *Scleronephthya*. Sinai, Red Sea. *Photo: KF*

2: Divaricate colony from the GBR, ca 5 cm. *Photo: KF*

3: Glomerate colony from the GBR, ca 5 cm. *Photo: KF*

4 AND 5: Supporting bundles, ca 1 cm. GBR. *Photos: KF*

6: Umbellate colony, Darwin, north Australia. *Photo: Karen Gowlett-Holmes*

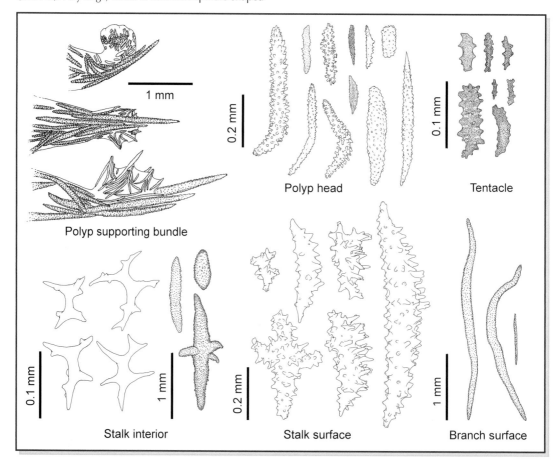

1 mm

0.2 mm

0.1 mm

Polyp head

Tentacle

Polyp supporting bundle

0.1 mm

1 mm

0.2 mm

1 mm

Stalk interior

Stalk surface

Branch surface

(Dendronephthya, continued)

Polyps: Monomorphic and not retractile. Tentacles are short and stout, with tips spread widely apart in disk-shaped arrangements when fully expanded. Polyps have a conspicuous supporting bundle of sclerites (see below).

Sclerites: The body of the majority of polyps is supported by a group of large spindle-shaped sclerites (several millimeters long) called a 'supporting bundle', which gives the colony a spiky appearance and feel. Similar spindles (often longer) occur in the surface of the twigs and branches. The polyp head carries numerous spindles arranged non-symmetrically in 8 points and sometimes a collaret. The tentacles contain small, flattened, irregular shaped forms. The upper part of the stalk surface generally contains thick spindles. The basal region tends to contain irregular-shaped sclerites, which often have thorny projections and may be branched. There are sometimes no sclerites in the stalk interior, but generally it contains spindles, sometimes branched, and/or minute antler-like forms. Sclerites are usually highly coloured.

Colour: Commonly bright colours, such as red, orange, purple, yellow, pink, or white, due to the colour of the sclerites. The sclerites in the stalk are often pale or white, while the bright colours occur in the upper branches. The same species may occur in several different colour forms. Azooxanthellate.

Habitat and abundance: Regionally common or even abundant below 20 m where currents are fast and unidirectional, but rare in wave-exposed areas. They can also grow in muddy estuaries and deep oceanic waters. In well-lit shallow water they are generally restricted to flow-exposed, wave-protected habitats or grow on steep walls. Although commonly occurring as individual colonies, a few species form extensive aggregations. Some multiply asexually by "polyp bail-out" (polyp bundles fall off the mother colony, settle, attach and grow up as new colonies; Photo 8).

Zoogeographic distribution: Widespread, from Africa to Micronesia and Polynesia.

Similar Indo-Pacific genera: Without close inspection, colonies of *Umbellulifera* and *Stereonephthya,* and even *Nephthea* and *Scleronephthya,* may be confused with this genus.

Photos:

7: Colonies in inter-reefal soft bottom areas of the Great Barrier Reef (GBR). *Photo: KF*

8: Recruits from "bail-out" of polyp bundles, which form "rhizoids", attach to the substrate and grow up as independent colonies. Northern Red Sea, ca 3 cm. *Photo: KF*

9 AND 11: Colonies with large and prominent supporting bundles. GBR, ca 8 cm. *Photos: KF*

10: A densely contracted glomerate colony, Hong Kong. *Photo: KF*

12: Intricate stalk surface sclerite arrangements, and polyp bundles. Northern Red Sea. *Photo: Günther Bludszuweit*

13 AND 14: On the GBR, colonies tend to be fully expanded at night. *Photos: KF*

7

8

Umbellulifera

Thomson & Dean, 1936

Colony shape: The characteristic shape of species within this genus is a very long, bare stalk, terminating in a single, relatively small, coloured, finely branched, umbellate polyp region. Rarely, colonies are found with other accessory polyp groups branching from lower down on the main stem. Contracted colonies up to half a meter tall are captured with deep water dredges, which would be well over a meter tall when expanded.

Polyps: Monomorphic, not retractile, and arranged singly, not in clumps, on the terminal twigs.

Sclerites: Polyp heads have numerous, small, stick-shaped sclerites with tall warts, symmetrically arranged in 8 points, but there are no collaret sclerites. Tentacle sclerites are irregular in shape, and may be branched or curved. Supporting bundles contain narrow spindles, usually with small warts. Supporting bundles are generally weak or absent altogether. Branches and stalk may be free of sclerites. If present, those in the surface of the branches can include asymmetrically developed capstans, short rods and branched forms with large warts, and long spindles with tall warts along one side. Those in the surface of the stalk are commonly asymmetrically developed capstans and irregular radiates, while those in the stalk interior are small, spiny, 6-radiates.

Colour: The polypary often contains brightly coloured sclerites, such as red, purple, pink, or yellow-brown. The stalk sclerites are sometimes colourless, but those in the upper portion are often coloured the same as those in the polyp region. Azooxanthellate.

Habitat and abundance: Rarely found, possibly because it grows mostly below diving depth.

Zoogeographic distribution: Not often recorded. Its occurrence in collections from the Red Sea, Madagascar, Indonesia, Philippines, Malaysia, Australia and New Caledonia indicate it can probably be found throughout the regions covered by this book.

Similar Indo-Pacific genera: Similar to any umbellate species of *Dendronephthya* with a long stalk.

Photos:

1, 2 AND **4:** Philippines. *Photos: Leen van Ofwegen*

3: Great Barrier Reef (GBR), ca 20 cm. *Photo: Queensland Museum*

5: A large dredged specimen. Rhizoids allow *Umbellulifera* colonies to grow in soft substrate. GBR, ca 40 cm. *Photo: Queensland Museum*

6: Sabah, Malaysia. *Photo: Frances Dipper*

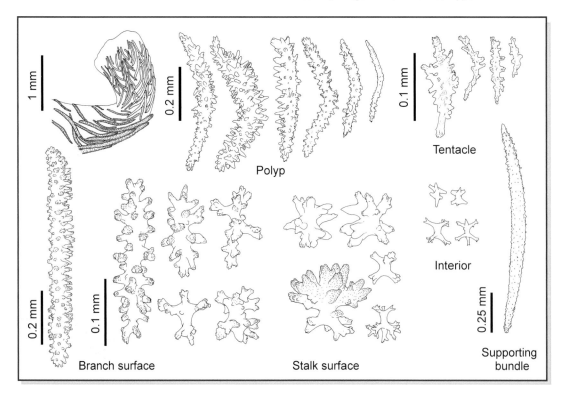

1 mm

0.2 mm

0.1 mm

Tentacle

Polyp

0.2 mm

0.1 mm

Interior

0.25 mm

Branch surface

Stalk surface

Supporting bundle

Leptophyton
van Ofwegen & Schleyer, 1997

and include small rods with spines and simple tubercles, and some crosses and capstan-like forms. Fewer and smoother versions of these forms are found in the surface of the branches, and also occur in eight groups in the lower parts of those polyps that grow from the colony base. Otherwise the polyps are without sclerites, and so is the colony interior.

Colony shape: Small, soft colonies with numerous branching lobes arising from a common base or a short stalk. Colonies look like a cluster of short lobes when contracted, but when they are expanded, the lobes become long, branch-like, and covered with polyps, giving them quite a bushy appearance.

Polyps: Small, monomorphic and retractile, though commonly preserved expanded. They are most abundant on the ends of the branches, but also occur on the common base, the major branches or lobes.

Sclerites: Sclerites are not present in large numbers. Those in the surface of the base are most numerous,

Colour: Pink, or white. Azooxanthellate

Habitat and abundance: Found on shallow overhangs or vertical walls of the coastal reefs at 18 - 30 m depths.

Zoogeographic distribution: *L. benayahui* is the only species described in this genus. It is only recorded from KwaZulu-Natal on the east coast of southern Africa.

Similar Indo-Pacific genera: May superficially resemble some *Stereonephthya*, *Scleronephthya*, or glomerate species of *Dendronephthya* when fully expanded, but those genera are generally rough to the touch.

Photos:
ABOVE AND BELOW: KwaZulu-Natal, South Africa. *Photos: Michael Schleyer*

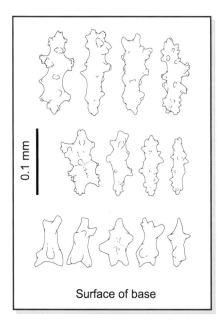

0.1 mm

Surface of base

Pacifiphyton
Williams, 1997

Colony shape: The colony shape is quite unique among soft corals. The stalk, which has a spreading holdfast, is stiff, brittle, very long (up to about 30 cm) and very thin. It is terminated by a star-like cluster of short branches and twigs that can be folded upwards like flower petals. These branches carry the polyps, which are more concentrated towards the ends.

Polyps: Monomorphic, not retractile, and not densely arranged.

Sclerites: The stalk gets its rigidity because it is constructed from long, very slender spindles, which are straight or sinuous, longitudinally arranged, and closely packed together. The branches are soft and flexible, and may contain a few similar spindles in the surface, or be sclerite free. The number and arrangement of sclerites in the polyps is also variable. The polyp body may contain just a few thin spindles (even just one) grouped on one side, or there may be a well-developed supporting bundle. The polyp head has a collaret and points arrangement, the tentacle backs may contain rods with a granular surface, and in the pinnules there are often small irregular shaped scales.

Colour: The stalk can be tan, light rose or flesh coloured, and the branches can be colourless, light

orange, or yellow-green and somewhat iridescent. The polyp body is colourless, the pharynx and mesenteries are apricot, and the tentacles are white.

Habitat and distribution: So far, it has been recorded only from 40 to 50 m water depths.

Zoogeographic distribution: *P. bollandi* is the only species described. So far, it has been recorded from Okinawa, Japan, and from Bali, Indonesia.

Similar Indo-Pacific genera: None.

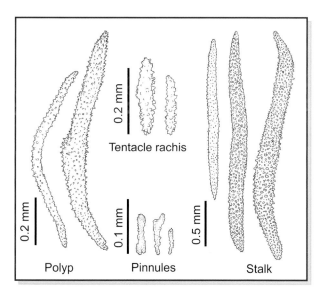

0.2 mm
Tentacle rachis
0.2 mm
0.1 mm
0.5 mm
Polyp   Pinnules   Stalk

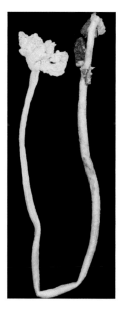

Photo:
ABOVE: Okinawa, Japan; ca 10 cm.
Photo: Robert Bolland
RIGHT: Preserved colony from Bali, Indonesia; total length: ca 15 cm.
Photo: PA

Lemnalia

Gray, 1868

Colony shape: Arborescent, slender colonies with conspicuously bare stalks and major branches. Branching numerous to sparse, with minor branches or twigs varying from long and thin to short and lobe-like. Individuals are not very contractile. The surfaces feel like fine sandpaper in preserved colonies. Colonies have an unpleasant odour, and when torn long mucus threads are usually apparent stretching across the broken section.

Polyps: Monomorphic, small, standing upright along minor branches and twigs. Not clustered in catkins or lobes, although terminal twigs can resemble catkins when the polyps are densely arranged. No supporting bundles. Polyps are very contractile, shrinking to small domes or cylinders when deflated (but see remarks).

Sclerites. The interior of the stalk, branches and twigs contains thin, relatively long sticks, which are ornamented with small prickles at the ends or all over. The surface of the narrow branches or twigs contain short sticks. The surface of the main branches and stalk generally contain several sclerite forms that include crescents and 6-radiate capstans, or modified forms of these. The tentacles commonly contain lobate scales, and the polyp body and head contains sticks and spindles, often curved, and sometimes with a pointed or club-like end. Colourless.

Colour: Cream to pale brown. Zooxanthellate.

Habitat and abundance: On the Great Barrier Reef, species are common on clear-water outer-shelf terraces at 10 - 25 m depths. Rare in areas of high wave energy and in turbid waters. Can be found as individuals or in small groups.

Zoogeographic distribution: *Lemnalia* probably occurs throughout most areas covered by this book. There are no records for Japan.

Similar Indo-Pacific genera: *Nephthea* and *Litophyton*. Some arborescent *Sinularia* also superficially resemble *Lemnalia* but are easily distinguished by closer inspection.

Remarks: Research has commenced to sort out some confusion amongst nominal species in this genus, as a couple of species have been reported to have retractile polyps. These species have lobe-like branches, thus linking the genus with the enigmatic species *Paralemnalia digitiformis*.

Photos:
1: Sabah, Malaysia, ca 15 cm. *Photo: Frances Dipper*
2 AND 3: Great Barrier Reef (GBR), ca 5 cm. *Photos: KF*
4 AND 5: GBR, ca 15 cm. *Photos: KF*
6: Zanzibar, ca 20 cm. *Photo: Matt Richmond*

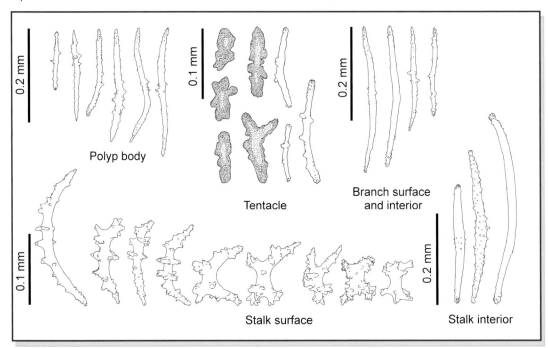

0.2 mm

Polyp body

0.1 mm

Tentacle

0.2 mm

Branch surface and interior

0.1 mm

Stalk surface

0.2 mm

Stalk interior

Paralemnalia

Kükenthal, 1913

Colony shape: Small and firm colonies consisting of a cluster of digitiform branches arising from a common base. Polyps occur only on the branches, sometimes only on the distal two thirds. Colonies have limited powers of contraction. They are brittle in the preserved state.

Polyps: Monomorphic, small, and retractile or non-retractile.

Sclerites: Polyp sclerites are very like those found in *Lemnalia*, but the spindles often have points at both ends. The interior sclerites are long, prickly spindles, also like those found in *Lemnalia*. The surface of the branches contains long spindles like those in the interior, along with shorter ones, often bow-shaped, which have several girdles of warts. These same sort of sclerites occur in the stalk surface along with short rods with girdles of spiny processes. Colourless.

Colour: Light brown or grey-brown, or cream with pink or brown polyps. Zooxanthellate.

Habitat and abundance: Common on flats and crests, and on terraces between 10 and 25 m depth on mid- and outer-shelf reefs of the Great Barrier Reef. Generally found in groups of only a few colonies. Rare on near-shore reefs.

Zoogeographic distribution: Red Sea, Madagascar, Indonesia, Australia, Philippines and Taiwan.

Similar Indo-Pacific genera: None, but see remarks.

Remarks: Research has commenced to sort out some confusion amongst nominal species in this genus, which at present contain species with retractile and non-retractile polyps. It should also be noted that *Paralemnalia digitiformis* and similar growth forms (eg, Photos 4 and 6), which are common on the Great Barrier Reef and which have retractile polyps, but robust and branched lobes, clearly do not fit the definition of this genus. These species have undeniable links to some enigmatic nominal species of *Lemnalia*.

Photos:

Below, and 5: Great Barrier Reef (GBR), 1 & 3 cm. *Photos: KF*

1: Zanzibar, ca 20 cm. *Photo: Matt Richmond*

2: Sabah, Malaysia. *Photo: Frances Dipper*

3: Papua New Guinea. *Photo: Coral Reef Research Foundation*

4: *Paralemnalia digitiformis*, GBR, ca 15 cm. *Photo: KF*

6: An atypical colony from the GBR, with retractile polyps, ca 5 cm. *Photo: KF*

7: GBR, ca 15 cm. *Photo: KF*

8: Palau, ca 8 cm. *Photo: KF*

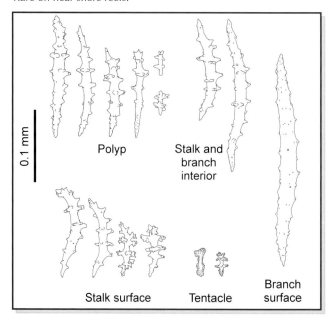

Polyp

Stalk and branch interior

Stalk surface

Tentacle

Branch surface

0.1 mm

Capnella
Gray, 1869

Colony shape: Colonies are small, usually less than 10 cm across, and generally lobate to weakly arborescent with polyps restricted to the lobes or branches. Most species have limited powers of contraction.

Polyps: Monomorphic, small, and contractile, with short tentacles. When they contract they become somewhat club-like, bending the head towards the colony surface and lying tightly against the lobes. Being closely packed together and covered in spiny sclerites they often give the lobes the appearance of a pine-cone or piece of scaly reptile skin.

Sclerites: The outer side of the polyps is densely packed with spiny or leafy clubs and spindles. The colony surface contains leafy or spiny capstans, and the interior is densely packed with warty ovals or spheres. Colourless.

Colour: Grey, beige or brown alive, commonly dark grey when preserved. Zooxanthellate.

Habitat and abundance: Common on crests and flats of mid- and outer-shelf reefs of the Great Barrier Reef, and on slopes down to 25 m depth. Some arborescent taxa prefer sheltered lagoonal conditions.

Zoogeographic distribution: Zanzibar, Madagascar, Seychelles, Sri Lanka, Indonesia, Australia and the Philippines.

Similar Indo-Pacific genera: Some arborescent *Capnella* may be mistaken for *Lemnalia, Nephthea* or *Litophyton.*

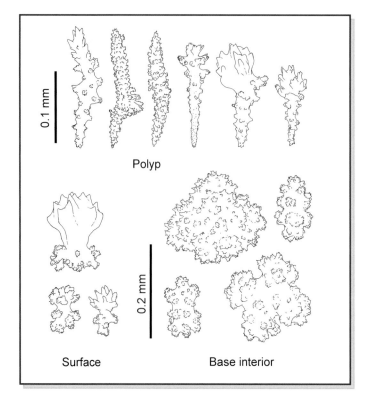

0.1 mm

Polyp

0.2 mm

Surface   Base interior

Photos:
ABOVE, **3, 4,** AND **6:** Close-ups of colonies from the Great Barrier Reef (GBR), views ca 3 - 4 cm. *Photos: KF*
1: GBR, view ca 20 cm. *Photo: KF*
2: GBR, view ca 10 cm. *Photo: KF*
5: Papua New Guinea. *Photo: Coral Reef Research Foundation*

Nidalia

Gray, 1834

Colony shape: Small, torch-like colonies with a sterile stalk and a hemispherical to globular capitulum which is sparsely or densely covered in large calyces. Contracted colonies are hard and rough to touch.

Polyps: Monomorphic, completely retractile, and relatively robust with a long body.

Sclerites: The calyces, and the surface of the capitulum and the stalk, contain large complexly warted spindles (which are occasionally branched). The interior of the colonies contains spindles of a similar architecture which are usually shorter and thinner than those of the surface. On the polyp head, curved spindles (often extremely numerous) occur in a collaret and points arrangement. The tentacles are generally densely packed with sclerites of many forms including rods, scales and stars. The entire polyp body, or just the basal portion, may contain narrow spindles or short rods and figure-eight scales. Sclerites are generally colourless or brown-orange. One species has red scales at the base of the polyp body.

Colour: Brown to pale orange, or yellow. Azooxanthellate.

Habitat and distribution: *Nidalia* appears to be a rare genus, and only 2 species are recorded to occur in less than 30 m of water.

Zoogeographic distribution: Shallow-water species are recorded from the Bay of Bengal, Timor, Bali, Papua New Guinea, Palau, and the Philippines.

Similar Indo-Pacific genera: *Bellonella*.

Photos:

1: Colonies from New Guinea, with retracted polyps. *Photo: Gary Williams*

2 AND 3: Beautifully expanded colonies, ca 4 cm. *Photo: Bert Hoeksema*

4: Malaysia, ca 4 cm. *Photo: Harry Erhardt*

5: *Nidalia simpsoni*, Indonesia, ca 4 cm. *Photo: Harry Erhardt*

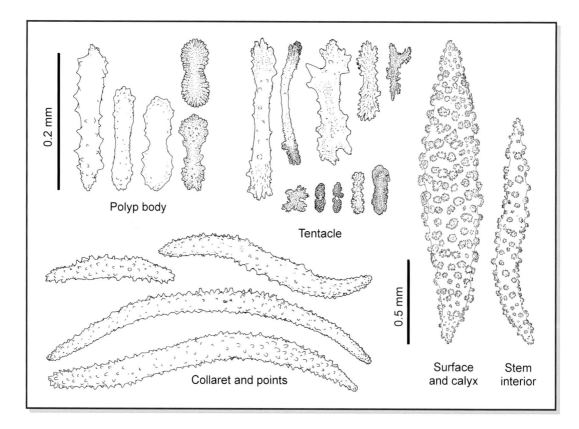

0.2 mm

Polyp body

Tentacle

Collaret and points

0.5 mm

Surface and calyx

Stem interior

Siphonogorgia

Kölliker, 1874

Colony shape: Colonies generally grow in one plane, profusely branched, and look just like gorgonians with a solid axis. Unlike gorgonians, however, there is no differentiated central axis, and they are easily broken. The main stem and main branches are rigid because they are formed by a thick layer of spindle-shaped sclerites densely packed around some central canals. The finer branches are more flexible because the spindle layer is much thinner, and the canals are well developed enabling the branches to inflate. In some species even major branches are inflatable. Polyps may occur on most of the branches, or they may be restricted to just the terminal branchlets, which can be lobe-like.

Polyps: Monomorphic, quite large, and able to completely retract within low calyces, although often preserved incompletely retracted. The calyces, which are formed from very small spindles and rods, are more obvious on the thinner branches where they are squeezed between the larger sclerites. On the thicker branches they may be reduced to a very narrow rim of a small number of sclerites around a conspicuous aperture.

Sclerites: The polyp head contains numerous large sclerites in a strong collaret and points arrangement. The polyp body below the head often contains numerous small rods, as does the tubular pharynx inside the polyp. The tentacles contain broad, flattened rods, and narrow, curved scales. The outer layer of the stem and all branches usually contains large, complexly warted spindles, narrow or robust, with shorter, thinner spindles and rods between them here and there. Sometimes in the terminal branchlets the outer layer only contains short sclerites. The walls of the internal canals contain short, narrow spindles and rods. Sclerites usually highly coloured.

Colour: Colonies are generally shades of brown, red, or dark purple, and polyps are commonly yellow or orange. Azooxanthellate.

Habitat and abundance: Uncommon but not rare. Generally found in clear-water habitats in wave-protected areas with good current.

Zoogeographical distribution: Probably occurs in most regions covered by this book, though confusion with *Chironephthya* in most of the literature makes this difficult to substantiate.

Similar Indo-Pacific genera: *Chironephthya*. Externally, colonies could also be confused with *Astrogorgia* in which the sclerites are spindles and commonly large, but *Siphonogorgia* does not have a horny central axis.

Photos:

1, 2 AND 8: Great Barrier Reef (GBR). *Photos: KF*

3 AND 4: Northern Red Sea, ca 10 cm. *Photos: Mark Wunsch*

5: Sabah, Malaysia, ca 12 cm. *Photo: Frances Dipper*

6 AND 7: The somewhat atypical species *S. geodeffry*, with expanded and retracted polyps. GBR, ca 8 cm. *Photos: KF*

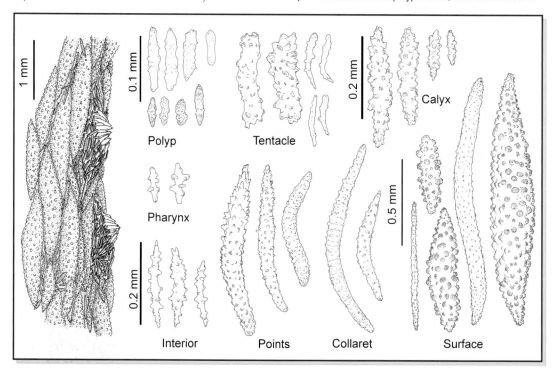

Polyp · Tentacle · Calyx · Pharynx · Interior · Points · Collaret · Surface

Chironephthya

Wright & Studer, 1889

Colony shape: Colonies are usually profusely branched. They may grow in one plane or they may be very bushy, and could be confused with gorgonians that have a solid axis. Unlike those gorgonians, however, there is no differentiated central axis, and colonies are easily broken. In the larger species, the stem and the main branches are rigid and uncompressible, because they are formed by a thick layer of spindle-shaped sclerites, densely packed around some central canals. In most smaller species, however, the stem and main branches are almost hollow, because the outer layer of spindles is relatively thin and surrounds a number of very wide canals. In both forms, the finer branches are flexible because the spindle layer is thin, and the canals are wide and well developed. Polyps are generally found only on the terminal and near terminal branchlets, though in some species they can be found on the main branches, and occasionally on the upper stem.

Polyps: Monomorphic, quite large, and able to completely retract within prominent calyces, although they are often preserved incompletely retracted. The calyces are of unusual construction (see below).

Sclerites: The calyces are shelf-like, growing on three sides of the polyp only, and are formed from numerous long spindles that are split in to two groups, like pointed ears. The polyp head contains numerous large sclerites in a strong crown and points arrangement. The polyp body below the head often contains numerous small rods, as does the tubular pharynx inside the polyp. The tentacles have broad, flattened rods, and narrow, curved scales. The outer layer of the stem and all branches contains large, complexly warted spindles, narrow or robust, with shorter, thinner spindles and rods between them here and there. The walls of the internal canals have short, narrow spindles and rods. Sclerites are usually highly coloured.

Colour: Shades of pink, white, red, yellow, brown or dark purple; often bi-coloured. The polyps are commonly yellow, white or orange. Azooxanthellate.

Habitat and abundance: Uncommon, generally found on walls and under ledges.

Similar Indo-Pacific genera: *Siphonogorgia*. Small colonies can also resemble *Nephthyigorgia*, but are easily distinguished by the structure of the calyces.

Zoogeographical distribution: Probably occurs in most regions covered by this book, though confusion with *Nephthyigorgia* and *Siphonogorgia* in most of the literature makes this difficult to substantiate.

Photos:

1: Ca 15 cm. *Photo: Bioquatic Photo - Alf J. Nilsen*

2, 6 AND **7:** Palau, ca 15 cm. *Photos: KF*

3 AND **4:** Papua New Guinea. *Photos: Roger Steene*

5: The characteristic split calyces are sometimes visible with unaided eyes. *Photo: PA*

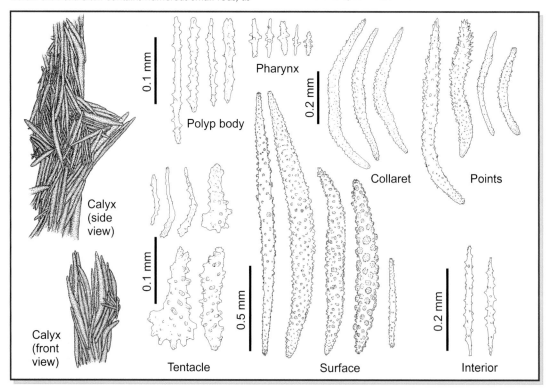

Calyx (side view)

Calyx (front view)

0.1 mm

Polyp body

Pharynx

0.2 mm

Collaret

Points

0.1 mm

Tentacle

0.5 mm

Surface

0.2 mm

Interior

Nephthyigorgia
Kükenthal, 1910

Colony shape: Small, firm, lobate or digitate colonies, commonly with a short stalk and often branched in one plane. Polyp calyces are large and conspicuous and confined to the lobes. Colonies are strongly contractile, and rough to the touch.

Polyps: Monomorphic, and retractile into prominent calyces. The calyces are usually volcano-shaped, or shelf-like where the upper side of the 'volcano' is reduced or absent.

Sclerites: The polyps are armed with a conspicuous collaret and points arrangement of slightly curved spindles. The spindles in the points may or may not have spiny tips. The rest of the colonial sclerites are all large, complexly warted spindles. Those in the calyces are longitudinally arranged, as are many of those in the dense colony surface layer. Those in the interior are sparse, and scattered longitudinally in the canal walls. Can be coloured.

Colour: Red, yellow, or white. Azooxanthellate.

Habitat and abundance: Rare in coral reef habitats, and usually found on muddy bottoms. Colonies generally occur as individuals.

Zoogeographic distribution: Very few records exist. Southern India in the Bay of Bengal, South China Sea, Indonesia, northern Australia, and the Great Barrier Reef.

Similar Indo-Pacific genera: *Chironephthya.*

Photos:

1: Vanuatu. *Photo: Harry Erhardt*

2: Colony kept in an aquarium. *Photo: Bioquatic Photo - Alf J. Nilsen*

3: Colony kept in a tank. Hong Kong, ca 10 cm. *Photo: KF*

4: Bali, ca 8 cm. *Photo: KF*

5 - 7: Great Barrier Reef. *Photo: Marine Bioproducts Group, AIMS*

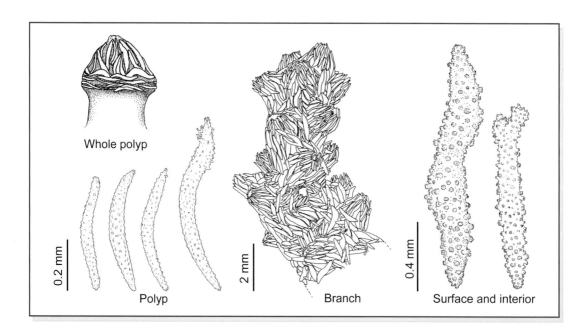

Whole polyp

0.2 mm

Polyp

2 mm

Branch

0.4 mm

Surface and interior

Studeriotes

Thomson & Simpson, 1909

Colony shape: Colonies have two distinctly different parts. The upper part consists of a soft, branched polypary with finger-like lobes, on which the polyps are arranged in small groups or lines. The basal portion of the colony is a hollow container-like cup, which can measure up to at least 20 cm in height and 10 cm in diameter. It has stiff walls, and is scarcely visible as most of it is usually buried in the sand or mud. The highly contractile upper portion of the colony is able to deflate and withdraw completely within this base, which then closes above.

Polyps: Monomorphic, small, with a supporting bundle.

Sclerites: The sclerites of the basal cup are very large, spiny spindles, sometimes branched, and up to 10 mm in length. There is a group of large spindles forming a supporting bundle for each polyp body, and each polyp head carries small spindles which may be arranged in eight double rows. Scattered spindles may be found in the surface of the lobes. Sclerites are colourless.

Colour: The upper, polyp-bearing region is generally whitish to pale brown. The polyps are darker coloured, purple, brown, or orange. Azooxanthellate.

Habitat and abundance: Not often reported, perhaps because it grows in muddy or sandy substrates where the water may be turbid.

Zoogeographic distribution: Singapore, Indonesia, Philippines, South China Sea, Andaman Islands, Taiwan, Palau, western, northern and eastern Australia.

Similar Indo-Pacific genera: The upper polyp-bearing portion could possibly be mistaken for *Lemnalia, Nephthea* or *Litophyton*.

Photos:

1: Southern Great Barrier Reef (GBR), ca 10 cm. *Photo: Gordon LaPraik*

2: Central GBR, ca 15 cm. *Photo: KF*

3: The basal cups of *Studeriotes* (13 and 8 cm long) attached to sea floor material. *Photo: PA*

4: *Studeriotes*, reared in a reef tank. *Photo: Bioquatic Photo - Alf J. Nilsen*

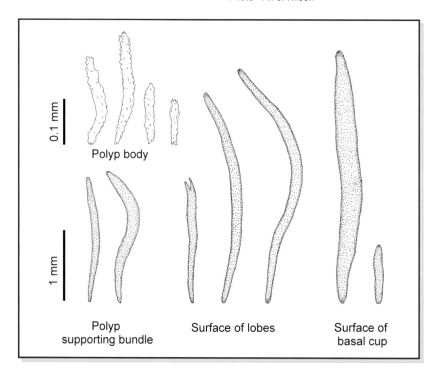

0.1 mm

Polyp body

1 mm

Polyp
supporting bundle

Surface of lobes

Surface of
basal cup

Asterospicularia

Utinomi, 1951

Colony shape: Colonies are soft and small, with a short stalk surmounted by several low rounded lobes of very regular shape and size (about 2 cm in diameter). When expanded, the short polyps make the lobes appear fluffy. Colonies have very limited powers of contraction.

Polyps: Monomorphic, contractile, small, with very short bodies (about 2 mm). The tentacles form cups when expanded. There are two nominal species, *A. laurae* Utinomi, 1951 and *A. randalli* Gawel, 1976. *A. laurae* was described as having no pinnules on the tentacles, however this is just a function of the extent of polyp contraction.

Sclerites: The colonies are densely filled with very small, stellate sclerites. The polyps generally contain minute disc or corpuscle-like sclerites. *A. laurae* was described as having no disc-like sclerites in the polyps, but the number of these sclerites varies enormously from polyp to polyp and colony to colony, and they are sometimes completely absent. Colourless.

Colour: Creamy with blue-green-opalescence, to creamy-brown. Colony and polyps the same colour, however the blue-green iridescence is restricted to the tentacles. Zooxanthellate.

Habitat and abundance: Moderately common in well-lit water on outer- and mid-shelf reefs of the Great Barrier Reef. They are often found in clones of tens of evenly sized, roundish colonies.
Common on reef flats and crests in wave-exposed habitats. In clear-water its distribution ranges down to 25 m depth, on less clear mid-shelf reefs it ranges to 10 m. It is absent on turbid coastal reefs.

Zoogeographic distribution: Taiwan, Guam, Palau, Papua New Guinea and the Great Barrier Reef.

Similar Indo-Pacific genera: Could be mistaken for a *Xenia* with very short polyps. *Sympodium* has similarly small polyps.

Photos:
BELOW, 2 AND 3: Great Barrier Reef, view ca 2 - 5 cm. *Photos: KF*
1, 4 AND 5: Great Barrier Reef, view ca 10 - 15 cm. *Photos: KF*

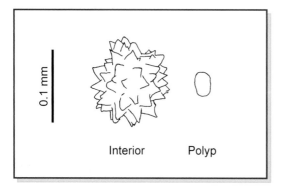

0.1 mm

Interior   Polyp

Xenia

Lamarck, 1816

Colony shape: Small, cylindrical or clavate colonies (typically 1 - 4 cm in diameter, 1 - 4 cm tall), sometimes branched, with dome-shaped summits. Polyps occur exclusively on the upper surface of the summits.

Polyps: Monomorphic, with varying contractility but never retractile. Tentacles relatively long, with 1 - 6 rows of pinnules arranged along both edges. The number of rows, and the number of pinnules per row are indicative of different species, but may vary with the age of the polyps or colonies, and the expansion of the tentacles. Polyps may or may not pulsate.

Sclerites: Minute, corpuscle-like platelets or spheroids, 0.02 - 0.05 mm diameter, with a fine granular surface. White, but iridescent. In some species sclerites are absent.

Colour: Colony stalks often coloured similar to the polyps, whitish, cream, yellow-brown, to dark brown. The sclerites in the polyps and the surface of the colonies often appear bright and opalescent. Zooxanthellate.

Habitat and abundance: Common in clear-water habitats from the reef flat to about 20 m depth where slopes are not too steep. Locally abundant in wave-exposed or current-swept areas. Often found in large groups containing tens of colonies. On coastal near-shore reefs, *Xenia* is rare and occurs patchily, often only below 10 m depth. It is absent in dirty and turbid water.

Zoogeographic distribution: Common in most regions from the tropical east coast of Africa to the western and central Pacific, with eastern limits being Melanesia and Palau, a northern extension to Japan, and south to the southern end of the Great Barrier Reef. Species of *Xenia* are the most abundant soft corals in many parts of the northern Red Sea.

Similar Indo-Pacific genera: *Heteroxenia* and *Asterospicularia.*

Photos:

BELOW: The extensions of this 8 x 10 m *Xenia* carpet hardly changed over a period of ~ 5 years, independently of the various types of neighbours living along the border line. Davies Reef, central GBR. *Photo: KF*

BELOW LEFT: The ca 0.02 mm sized sclerites of *Xenia*. *Photo: PA*

1: Papua New Guinea. *Photo: Roger Steene*

2 AND 5: Sabah, Malaysia, ca 5 cm. *Photos: Frances Dipper*

3, 4 AND 6: Great Barrier Reef (GBR), 5 - 15 cm. *Photos: KF*

7: Colony contraction at noon in this species is probably induced by high levels of irradiance. Northern Red Sea. *Photo: KF*

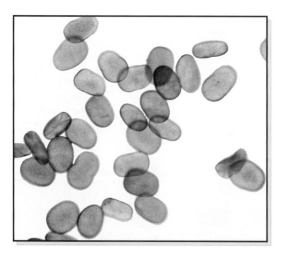

Heteroxenia
Kölliker, 1874

Colony shape: Cylindrical or clavate colonies, relatively small (2 - 7 cm in diameter), sometimes branched, with dome-shaped summits. Polyps occur exclusively on the upper surface of the summits.

Polyps: Mature colonies are dimorphic, having numerous, small siphonozooids nestled between the bases of the large autozooids. There is conflicting evidence as to whether the appearance of siphonozooids is seasonal. Prof. Y. Benayahu (pers. com.) says juvenile colonies in the Red Sea only have autozooids, developing siphonozooids when they become sexually mature. Gohar (1940) reported that the siphonozooids were only present during the breeding season, and cited *H. fuscescens* as an example. But Prof. Benayahu says that in over 20 years of research on *H. fuscescens* he has never seen this phenomenon. Dr G. Reinicke, however, reports seeing seasonal siphonozooids in other species of the genus (pers. com.).

Autozooids vary in their extent of contractility, but are never retractile. The tentacles are relatively long, with 1 - 5 rows of pinnules arranged along both edges. Polyps may or may not pulsate. The siphonozooids are small, flush with or slightly raised above the colony surface, with the tentacles reduced to mere knobs.

Sclerites: Minute corpuscle-like platelets or spheroids about 0.02 mm diameter, with a fine granular surface. White, and only slightly iridescent. In some species sclerites are absent.

Colour: Autozooids are white or cream, the long bodies often with faint bands. The colony surface is whitish, occasionally light brown, and the same colour as, or lighter than, the polyps. The sclerites in the polyps and the surface of the colonies often appear opalescent. Zooxanthellate.

Habitat and abundance: Often found as individual colonies; large aggregations are rarely encountered. They are uncommon on the Great Barrier Reef, but individual large colonies or clones of small colonies can be found on near-shore reefs below 10 m, or in shallow, lagoonal or back-reef environments of clear-water reefs. Very abundant in the northern Red Sea.

Zoogeographic distribution: Wide-spread. Red Sea, Madagascar, Mozambique, Indonesia, the western parts of Micronesia, Philippines, Taiwan, Fiji, northwest Australia, Great Barrier Reef.

Similar Indo-Pacific genera: When in their monomorphic phase, *Heteroxenia* species look just like *Xenia*.

Photos:

BELOW, AND 5: Many, but not all *Heteroxenia* have pulsating autozooids. GBR. *Photos: KF*

BELOW LEFT: The about 0.02 mm sized sclerites of *Heteroxenia*. Photo: PA

1: Colony with sparse, large autozooids in amongst a dense layer of siphonozooids. Great Barrier Reef (GBR), ca 8 cm. *Photo: KF*

2: Some *Heteroxenia*, in contrast to many other xeniids, can live in muddy environments. GBR, ca 10 cm. *Photo: KF*

3: Colony from Sabah, Malaysia, ca 5 cm. *Photo: Frances Dipper*

4: Siphonozooids are often only visible in contracted colonies. GBR, ca 12 cm. *Photo: KF*

1

2

3

4

5

Funginus

Tixier-Durivault, 1978

This genus appears to be the same as *Heteroxenia* – see Remarks.

Colony shape: Small, cylindrical or clavate colonies with polyps restricted to a dome-shaped summit.

Polyps: Dimorphic. Autozooids are few and very contractile, deflating until virtually flush with the colony surface, in which instance the pinnules become so small they may be difficult to detect. The siphonozooids are distributed more towards the centre of the summit.

Sclerites: Minute, corpuscle-like platelets or spheroids, about 0.02 mm diameter, with a fine granular surface. White, and only slightly iridescent.

Colour: Unknown.

Habitat and distribution: Very rare, only two specimens appear in the literature.

Zoogeographic distribution: New Caledonia and the Ryukyu Islands, from published accounts. A specimen from the Great Barrier Reef is housed at the Museum and Art Gallery of the Northern Territory, Australia.

Similar Indo-Pacific genera: *Heteroxenia, Xenia.*

Remarks: The only species known is *Funginus heimi*, originally from New Caledonia, and described as having retractile autozooids, and tentacles without pinnules. We have seen this specimen, and although what autozooids remain are highly deflated and contracted, several tentacles are visible and the pinnules can be detected as small bumps (about 7 down each side of a tentacle). Notably, also, there is no sign of polyp retraction at all, as was claimed in the original description. The autozooids appear to be only partly formed, with few, if any, tentacles, and seem to be regenerating, presumably after predation. There is also a portion of the colony summit where neither autozooids nor siphonozooids are present, which supports the predation theory.

Half of the summit of a specimen from the Great Barrier Reef, which we have illustrated, has the same appearance as the New Caledonian *Funginus* specimen. The autozooids in this area are clearly regenerating, several having only a single tentacle remaining.

The microstructure of the sclerites of *Funginus* and *Heteroxenia* is identical, and although more material needs to be examined, we feel *Funginus heimi* will prove to be just another species of *Heteroxenia*.

This genus was originally called *Fungulus*, but that name had already been used for an ascidian, and the new name *Funginus* was substituted in 1987 (Tixier-Durivault 1987).

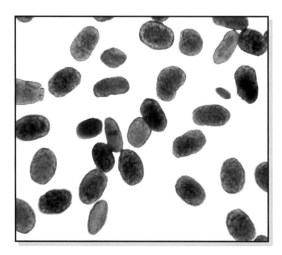

Photos:

LEFT: The ca 0.02 mm sized sclerites of the specimen housed at the Museum and Art Gallery of the Northern Territory in Darwin. *Photo: PA*

1: The museum specimen housed at the Museum and Art Gallery of the Northern Territory in Darwin. Colony size: ca 2.5 cm. *Photo: PA*

2 AND 3: Close-ups of the above colony, showing clearly the existence of pinnules in the autozooids, and the unusually large and well-developed siphonozooids. *Photos: PA*

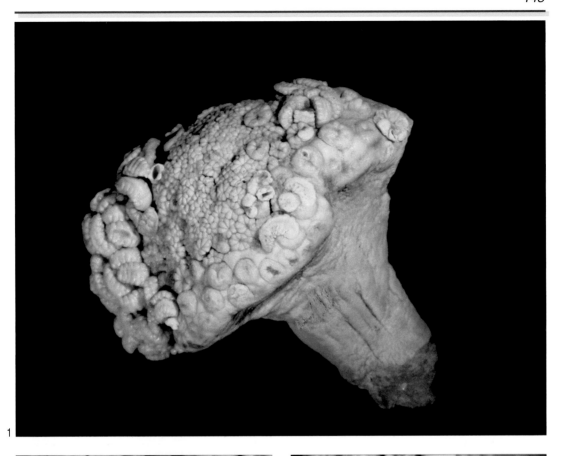

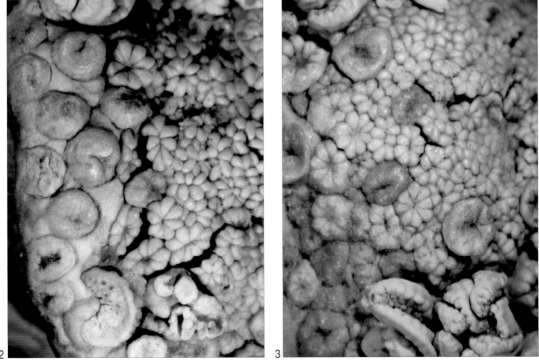

Efflatounaria

Gohar, 1934

Colony shape: Colonies generally consist of a number of lobe-like branches, arising from a short stalk. Some colonies appear as a close cluster of lobes joined at their base. The polyps are restricted to the lobes and the upper part of the stalk. Colonies often reproduce asexually, and it is very common to see individual colonies either joined by stolons, or sending out stolons from the colony base or from the tips of long, snake-like branches. Daughter colonies bud from the stolons, which later become re-absorbed.

Polyps: Monomorphic, of medium size, and with a relatively short body. They are so highly contractile that they can deflate until flush with the colony surface and appear to be retracted. The tentacles have one row of 4 - 6 pinnules along each side. Polyps occur on the lobes and the upper stalk, their size decreasing towards the basal regions of the colony, and they do not pulsate.

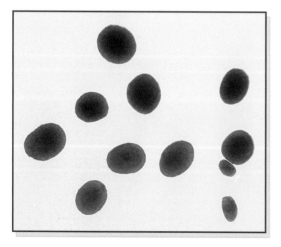

Sclerites: Minute, corpuscle-like platelets or spheroids, about 0.02 – 0.12 mm diameter, with a fine granular surface. White, but iridescent. In at least one species sclerites are absent.

Colour: Cream, white, bluish, yellow, greenish-yellow, yellow-brown, or brown. The colony surface has generally the same colour as the polyps, but the interior of the colony can be yellow. Zooxanthellate.

Habitat and abundance: Very abundant to dominant on exposed outer-shelf reefs of the Great Barrier Reef, predominantly on terraces below 5 m. Here, white, pink or bluish forms are often found in extensive clones of tens of colonies. A yellow-brown, soft, *Klyxum*-like species with long slender branches is locally common in turbid near-shore waters away from river estuaries. Otherwise, since few records exist, the genus appears to be uncommon throughout the Indo-Pacific.

Zoogeographic distribution: South Africa, Tanzania, Madagascar, Philippines, Great Barrier Reef. Not recorded from the Red Sea, Guam, Palau, and the eastern Pacific.

Similar Indo-Pacific genera: Can be confused with *Cespitularia*, but *Efflatounaria* is stouter and has highly contractile polyps. A coastal species resembles *Klyxum*.

Remarks: When Gohar established the genus *Efflatounaria* in 1934, he characterised it as having retractile polyps. In actual fact the polyps are just very highly contractile, and give the appearance of being retracted when fully deflated. The genus is based on the species *E. tottoni*, which did not have sclerites, but at the same time Gohar included in the genus a species with sclerites.

Photos:

LEFT: The 0.02 - 0.12 mm sclerites of *Efflatounaria*. *Photo: PA*

ABOVE LEFT, AND 1: *Efflatounaria* on outer-shelf reefs of the Great Barrier Reef (GBR). *Photos: KF*

2 AND 5: A brown, translucent coastal species with polyp-free branch tips which act as stolons. GBR, 5 - 10 cm. *Photo: KF*

3: Characteristic for *Efflatounaria* are the contractile polyps. GBR, ca 4 cm. *Photo: KF*

4: A yellow off-shore species. GBR, ca 10 cm. *Photo: KF*

6: Stolons contribute to the ability to rapidly colonise dead substrate. Sabah, Malaysia. *Photo: Frances Dipper*

7: *Efflatounaria* on the Great Barrier Reef are commonly found with some of their branches bitten off by an unknown predator. *Photo: KF*

Cespitularia

Milne-Edwards & Haime, 1850

Colony shape: Colonies are very soft, and somewhat transparent, with a few lobe-like branches. Polyps occur only on the branches and upper parts of the stalk.

Polyps: Polyps are like those of *Xenia*, but are not restricted to the branch ends. They are monomorphic, relatively long, non-retractile and only slightly contractile. The tentacles are often held in a basket-shaped arrangement, with tips close together, and often pulsate. Pinnules are usually very numerous, and arranged in 1 - 4 rows along both edges of each tentacle.

Sclerites: Minute, corpuscle-like platelets or spheroids, 0.02 – 0.04 mm diameter, with a fine granular surface. White, but iridescent. In some species sclerites are absent.

Colour: Colony surface white, cream, blue, iridescent-green or pale brown. Polyps have the same colour or are darker than the surface, in particular the tentacles which are often brown. Sometimes the pinnules are coloured darker than tentacles and branches, giving the polyps a striped appearance. Zooxanthellate.

Habitat and abundance: On the Great Barrier Reef, they are fairly common on shallow wave-protected terraces, both in moderately turbid and clear-water environments. They are often found in groups of few to tens of colonies.

Zoogeographic distribution: Found throughout many of the regions covered in this book, but absent in Guam.

Similar Indo-Pacific genera: *Efflatounaria*, but in that genus the polyps can contract until they are flush with the colony surface.

Photos:

BELOW LEFT: The 0.02 mm sclerites of *Cespitularia. Photo: PA*

BELOW, AND 6: *Cespitularia* from the Great Barrier Reef (GBR) with their characteristic non-retractile polyps. *Photos: KF*

1: Green-iridescent colony from the northern GBR. *Photo: KF*

2 AND 3: Groups of colonies from the GBR. *Photos: KF*

4: Slender *Cespitularia* from Sabah, Malaysia, with unusually reduced tentacles. *Photo: Frances Dipper*

5: Blue-iridescent *Cespitularia* from Sabah, Malaysia. *Photo: Frances Dipper*

Sympodium
Ehrenberg, 1834

Colony shape: Colonies consist of a basal membrane or ribbon-like stolons, from which small polyps arise. Although the basal membrane is usually thin, it can be several millimeters thick, and parts of a colony can form irregularly shaped mounds. Some parts of the basal membrane may have few or no polyps, while in other sections they are very crowded.

Polyps: Monomorphic and completely retractile. They are very short, and when fully expanded give the colony a furry appearance. The number and arrangement of the pinnules are unknown.

Sclerites: Minute corpuscle-like platelets or spheroids, about 0.02 mm in diameter, with a fine granular surface. White, but iridescent.

Colour: The polyps are brown, and the basal membrane is bright white or blue-white, and barely visible when the polyps are fully expanded. The polyps leave small, dark pits when they retract. Zooxanthellate.

Habitat and abundance: Uncommon. On the Great Barrier Reef it is restricted to clear-water reefs. Patches usually consist of only a few colonies, and often only one.

Zoogeographic distribution: Red Sea, Madagascar, Okinawa, Philippines, Guam, Palau, and the Great Barrier Reef.

Similar Indo-Pacific genera: *Asterospicularia* and *Rhytisma* can have similar growth form and small polyps.

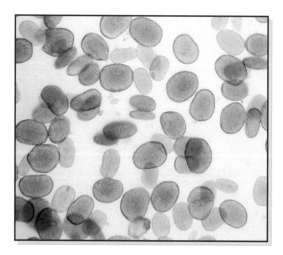

Photos:

LEFT: The 0.02 mm sclerites of *Sympodium*. *Photo: PA*

1: The characteristic change of colour from brown to white during polyp retraction is caught in this close-up image. Great Barrier Reef (GBR), ca 5 cm. *Photo: KF*

2 AND 3: GBR, 10 - 15 cm. *Photos: KF*

4: Colony with fully retracted polyps. GBR, ca 5 cm. *Photo: KF*

5: Expanded colony, GBR, ca 5 cm. *Photo: KF*

Sansibia

Alderslade, 2000

Colony shape: Small or large clusters of polyps, united basally by a thin, soft membrane or by ribbon-like stolons that encrust the substrate.

Polyps: Monomorphic, up to several centimeters tall, and slightly contractile, but not retractile. Tentacles have numerous pinnules arranged in 1 - 4 rows along both edges.

Sclerites: Minute corpuscle-like platelets or spheroids, about 0.02 mm in diameter, with a fine granular surface. White, but iridescent. There are probably species in which the sclerites are absent.

Colour: Brown, with areas of iridescent colours, usually green and/or blue, predominantly in the tentacles. Contains high densities of zooxanthellae.

Habitat and abundance: *Sansibia* seems to be uncommon, but only few data are available. Occasionally found in very turbid water. It has also been recorded from an intertidal mud-flat in Darwin, north Australia.

Zoogeographic distribution: Scattered records from warm, subtropical and tropical waters of the Indian and Pacific Oceans, and associated seas. Records span from the coast of Africa across to Hawaii in the west, and from Taiwan in the north down to the coast of northern New South Wales in Australia.

Similar Indo-Pacific genera: *Anthelia, Cervera, Stereosoma.*

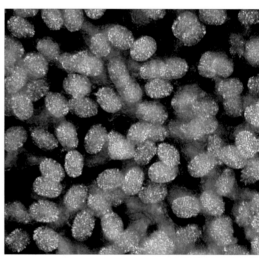

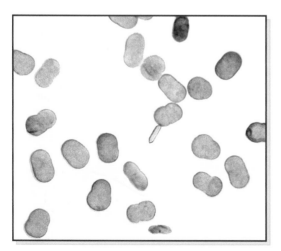

Photos:

ABOVE LEFT: Close-up of the mouth opening, showing the iridescent sclerites. Darwin, northern Australia, ca 0.5 cm. *Photo: PA*

ABOVE: The opalescent sclerites in a live *Sansibia* colony from Darwin, northern Australia. View ca 0.3 mm. *Photo: PA*

LEFT: The same sclerites as above, shown through a microscope. *Photo: PA*

1, 3, 6 AND 7: An intertidal species from Darwin, northern Australia, views 2 - 5 cm. *Photos: PA*

2 AND 5: Hong Kong. *Photos: KF*

4: Hawaii. *Photo: Kevin Reed*

Anthelia

Lamarck, 1816

Colony shape: Small clusters of polyps, united basally by a thin, soft membrane or by ribbon-like stolons that encrust the substrate.

Polyps: Monomorphic, up to several centimeters tall, slightly contractile but not retractile. They have a noticeably thick body, and numerous pinnules arranged in 1 - 4 rows along both edges of each tentacle. Some species do not respond to touch. With species on reef flats of the GBR, the expanded tentacles generally stretch no larger than 0.5 cm in diameter. Further down the slope they may be up to 2 - 3 cm in diameter.

Sclerites: Short, somewhat flattened rods, about 0.04 – 0.13 mm long, with a very coarse, crystalline surface structure often in a herring-bone pattern.

Colour: Polyps and membrane are generally the same colour, which is often white or beige due to a high concentration of sclerites. Unlike most other genera in this family, the sclerites in the polyps and the surface of the colonies do not appear bright and opalescent. Zooxanthellate.

Habitat and abundance: Locally moderately common. Patches consist of only few colonies, often only one. Some species appear restricted to reef flats, and others to reef slopes.

Zoogeographic distribution: Wide-spread. Recorded from the Red Sea, Madagascar, Mauritius, central Indo-Pacific, Malaysia, Philippines, Japan, Melanesia, Micronesia, Great Barrier Reef, New Caledonia.

Similar Indo-Pacific genera: *Sansibia*. The expanded polyps could also be confused with *Tubipora* and *Clavularia*, however the latter two genera respond to touch and retract.

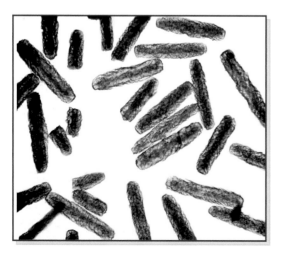

Photos:

ABOVE: *Anthelia*, with the characteristic membranous base that encrusts the substrate. Great Barrier Reef (GBR), ca 12 cm. *Photo: PA*

LEFT: The 0.04 - 0.13 mm sclerites of *Anthelia*. *Photo: PA*

1, 2 AND 7: Sabah, Malaysia, ca 10 cm. *Photos: Frances Dipper*

3 AND 8: Zanzibar, ca 10 cm. *Photos: Matt Richmond*

4 - 6: GBR, ca 10 cm. *Photos: KF*

Briareum

Blainville, 1830

Colony shape: Species may form thin encrusting membranous sheets, small clusters of knobs, tall finger-like lobes, or large tangles of cylindrical branches. In some species the branches and lobes have a hollow centre, and species that are membranous may appear to form lobes when they overgrow an irregular surface. In some species, younger parts of a colony may overgrow older parts, and successive layers may become linked by stoloniferous platforms, producing filo-pastry like structures. Polyps that are completely submerged under new layers generally become resorbed. The formation of new layers can occur as a response to excessive silting or overgrowth by sponges. Live foreign tissue that subsequently becomes enclosed may die, and bacterial degradation or subsequent burrowing by animals such as polychaete worms may result in cavities eventually becoming virtually free of debris. Large colonies may cover several square meters. In some species, the polyp-bearing part of the colony is virtually smooth, but most species have calyces. The calyces can vary from very low mounds (PHOTO 5) to tall, narrow structures 15 mm high (PHOTO 10).

Polyps: Monomorphic, retractile, and up to about 15 mm tall. The oral disk may protrude or it may be on the same level as the oral portion of the tentacle bases. Tentacles are variable in shape, ranging from thin to flattened, and in some species the pinnules are so reduced that they are barely visible. If the tentacles of an emerging polyp are observed through a strong magnifying lens, it generally becomes obvious that they have been invaginated like the finger of a glove that has been turned outside in.

(continued)

Photos:

1: Encrusting colony with beige surface, here with almost contracted polyps. Great Barrier Reef (GBR), 10 cm. *Photo: KF*

2: Encrusting species of *Briareum* with purple surface and large calyces. GBR, view ca 12 cm. *Photo: KF*

3: Colony with fully expanded polyps. Sabah, Malaysia, ca 8 cm. *Photo: Frances Dipper*

4: A branching species of *Briareum*. *Photo: Julian Sprung*

5 AND 6: Close-up of two colonies without conspicuous calyces. GBR, ca 3 cm. *Photos: KF*

7 - 10: Preserved colonies, demonstrating a range of variations in surface colouration and calyx sizes. View ca 3 cm. *Photos: PA*

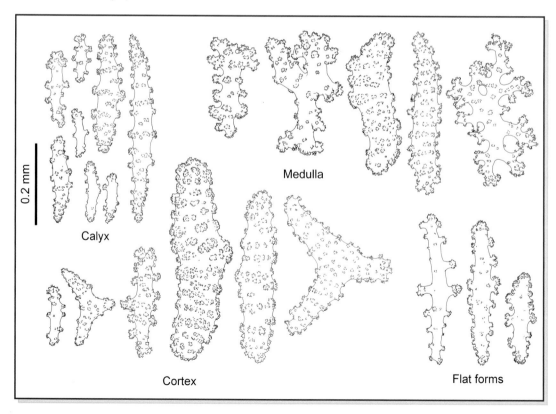

0.2 mm

Calyx

Medulla

Cortex

Flat forms

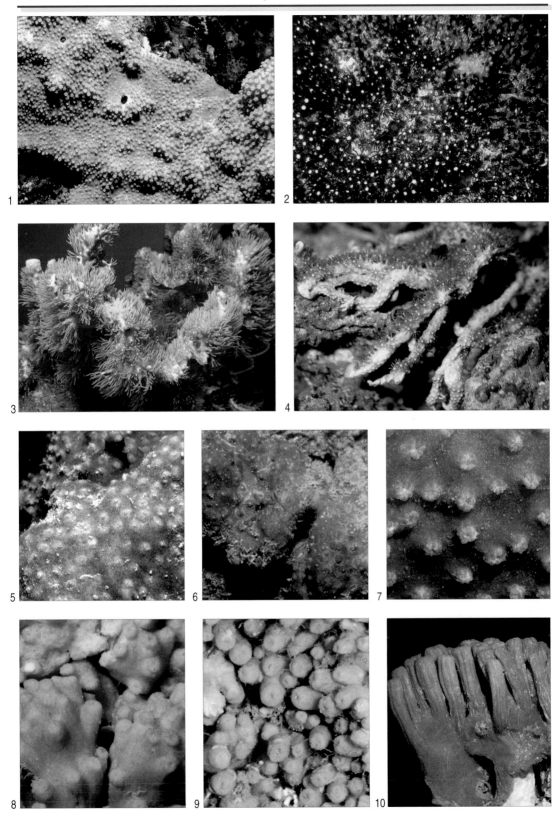

(Briareum, continued)

Sclerites: *Briareum* sclerite forms are quite unusual, and they are only found in this genus. With rare exceptions, all spindles and branched spindles have low or tall, spiny tubercles arranged in relatively distinct girdles, and the multiple-branched, reticulate and fused forms generally have very tall, complex tubercles. In rare cases the spindles are flattened, and the tubercles only occur along two sides. Polyps generally do not have sclerites, but in a few species there are small spindles around the base where it merges with the calyx rim. The summit of the calyx has small, sometimes flattened spindles around the polyp aperture, commonly arranged in 8 triangular groups. The rest of the calyx, if tall enough, contains long, narrow spindles, longitudinally arranged in the wall, sometimes forming 8 groups or ridges. The shorter the calyces, the fewer the number of long spindles, and in very low calyces these forms are absent. In the surface of the colony there is a thin layer of short, narrow spindles. Underneath this is a thicker layer of longer, generally robust spindles, occasionally branched. Below this is a basal layer, or medulla, which is almost always coloured deep magenta in Indo-Pacific species. The basal part of this layer usually contains mostly branched sclerites, often complex and reticulate, which may fuse into small clumps. If the coloured layer is thick, the upper part will contain long, generally robust, magenta spindles, occasionally branched. Sclerites are either white, or pale to deep magenta.

Colour: Layers of different coloured sclerites are a notable characteristic of this genus. The basal layer (called the medulla) is virtually always deep magenta, but magenta coloured sclerites may also be found in differing proportions in other parts of the colony. In live colonies, the upper or outer surface can vary from cream-brown to grey-brown to pink-brown to purple-brown to reddish-purple, and some parts of a colony may be coloured different to others. In preserved material, it becomes obvious that these colour differences can be attributable to the different proportions of magenta sclerites in the surface layer. This can vary from nil, or just a few of these coloured sclerites in the rim of the polyp aperture (which may be the tip of a calyx), to cases in which the whole colony surface is covered with them. The coenenchymal layer between the surface and the magenta medulla may contain just white or just magenta sclerites, or a mixture of the two. Zooxanthellate.

Habitat and abundance: Moderately common in a wide range of habitats, but tends to avoid very clear water. In very turbid water it may be found on highly illuminated surfaces. In clearer water it grows predominantly on vertical or shaded substrates. On illuminated upper surfaces it may suffer from bleaching from which it appears to recover only very slowly. On the Great Barrier Reef, this genus is common on near-shore reefs where it can cover extensive fields (hundreds of square meters), and is particularly common in the upper 8 m on fore-reefs.

Zoogeographic distribution: Records are few, but it has been reported from the Red Sea, east Africa, and the Indo-West-Pacific, including Australia, Indonesia, Micronesia, Taiwan and the Bonin Islands to the north. It also occurs in the Caribbean.

Similar Indo-Pacific genera: The Caribbean *Erythropodium* species has the same magenta basement layer as *Briareum*, but the only Indo-Pacific *Erythropodium* species grows as narrow, thin stolons. *Rhytisma* has a membranous growth form, but the colour and the sclerite shapes are quite different.

Remarks: The species with very tall, deep magenta-coloured calyces have usually been referred to in the literature as *Pachyclavularia* Roule, 1908. The production of cavities and platforms is prevalent in these long polyp forms, but is not restricted to them. They often grow in quite silty habitats, and the calyx height may be another adaptation to such an environment. The Indo-Pacific membranous and hollow-branched forms of *Briareum* are recorded in the older literature under the name *Solenopodium* Kükenthal, 1916. *Clavularia hamra* from the Red Sea is incorrectly named and should be called *Briareum hamra*.

Photos:

11: A preserved colony of *Briareum hamra* from the Red Sea. *Photo: PA*

12: A cross-section of a preserved colony from Okinawa, Japan. *Photo: PA*

13: Some coastal reefs on the central Great Barrier Reef are dominated by extensive carpets of *Briareum*. *Photo: KF*

14: A branching colony from the Solomon Islands. *Photo: PA*

15 AND 16: A branching colony from Rowley Shoals, NW Australia. *Photos: PA*

17 - 19: Close-ups of a range of polyp types of *Briareum*. Great Barrier Reef, 3 - 5 cm. *Photos: KF*

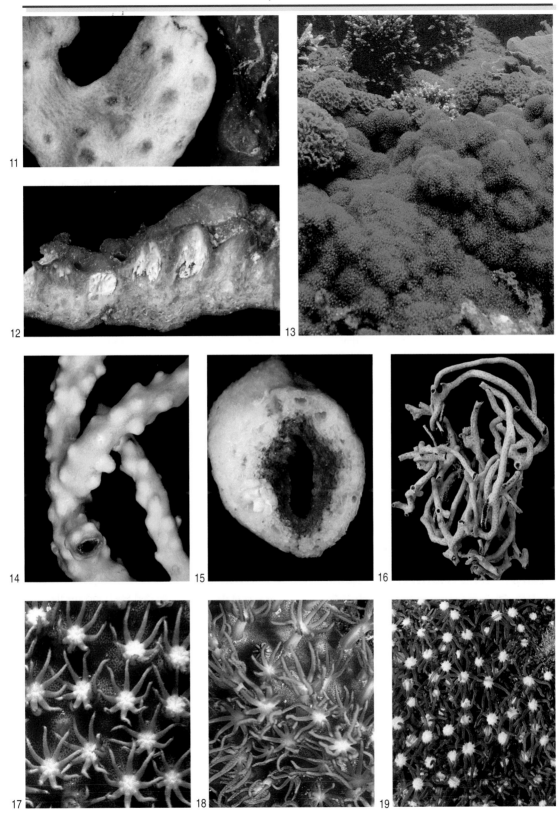

Iciligorgia

Duchassaing, 1870

Colony shape: Large colonies growing irregularly in one plane, carrying fragile, sometimes thick branches which do not form nets. The branch ends are often swollen, and each is grooved or has an elongate dish-like depression on the backside. Branches may be flattened and asymmetric in cross-section, due to the thickened cortex on the sides where the polyps are situated. The axis consists entirely of closely-packed sclerites, and breaks very easily.

Polyps: Monomorphic, of medium size, retractile into low hemispherical calyces, and crowded on three sides of the smaller branches, and the front of larger branches and the stem.

Sclerites: Those of the axial medulla are predominantly long needle-like forms with a few prickles. The outer layer (cortex) contains warty spindles and ovals. The cortex is missing from the bottom of the terminal grooves or dish-like depressions, revealing the medulla. The polyps have spindles in a collaret and points arrangement. Sclerites are colourless.

Colour: Branches yellow, brown, or dark red. Polyps white, or the same colour as the coenenchyme. Azooxanthellate.

Habitat and abundance: Generally restricted to muddy, silty environments.

Zoogeographic distribution: Indonesia, Philippines, Papua New Guinea, the Solomon Islands, northern Australia, Great Barrier Reef, and New Caledonia. Also found in the West Indies and other parts of the Caribbean.

Similar Indo-Pacific genera: *Solenocaulon.*

Remarks. In the Indo-Pacific this genus has been known as *Semperina*, but appears to be the same as the West Indian earlier named genus *Iciligorgia* (Grasshoff 1999). The genus *Solenocaulon* is very closely related, and essentially is only distinguished by the fact that the dish-like depressions in the ends of the terminal branches are exaggerated to such an extent that the end branches, and often also main branches, become tubular or gutter-like. Moreover, *Solenocaulon* but not *Iciligorgia* is often anchored in soft substrate.

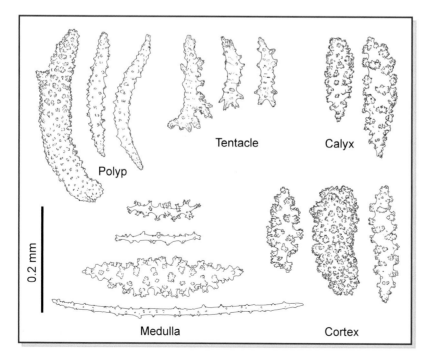

Polyp

Tentacle

Calyx

0.2 mm

Medulla

Cortex

Photos:

1 AND 4: Northern GBR, view ca 30 cm. *Photos: KF*

2 AND 3: Southern GBR. *Photos: Gordon LaPraik*

5: Close-up of *Iciligorgia* branches with their irregularly distributed polyp calyces, and clearly visible terminal grooves. Northern GBR. *Photo: KF*

6: Close-up of branch tips. Southern GBR. *Photo: Gordon LaPraik*

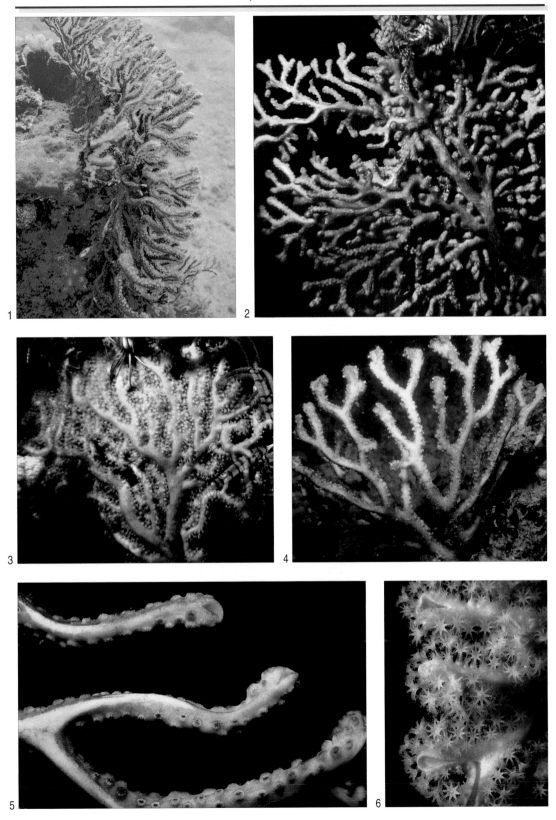

Solenocaulon

Gray, 1862

Colony shape: Colonies generally of moderate size (ca 15 cm; however a few very large colonies, with fans of 1.5 m diameter, have also been recorded), and irregularly branching in one plane. Although the stem and some major branches may be solid, most or all of the branches are hollow, tubular or gutter-like, and may have numerous large holes in their sides. The ends of tubular branches tend to have portions of the sidewalls flattening out in an irregular leaf-like manner. Colonies may attach to substrate including loose coral and shell fragments, but commonly have a flat, spatulate, basal extension that anchors them in soft substrate. Colonies are easily broken, as the axial substance that forms the centre of the stalk and the inner layer of the walls of the tubes, is formed only from closely packed sclerites.

Polyps: Monomorphic and retractile into low dome-like calyces. Their distribution varies amongst the species. They are commonly arranged irregularly along the rims of the holes and gutters, they may also occur densely or sparsely on branch surfaces, and they may be more or less confined to the smaller terminal branches.

Sclerites: Those of the axial medulla, which also forms the interior surface of tubes and gutters, are predominantly long needle-like forms with a few prickles. The outer cortex layer contains warty spindles and ovals. The polyps have spindles in a collaret and points arrangement. Sclerites are colourless.

Colour: Branches may be pink, white, yellow, brown, or red. Polyps are generally white. Azooxanthellate.

Habitat and abundance: Uncommon. Generally restricted to muddy, silty environments with strong currents and weak wave action.

Zoogeographic distribution: Most records are from the central Indo-Pacific, but also reported from the Persian Gulf, Zanzibar, Madagascar, Maldives, South China Sea, Taiwan, and the Great Barrier Reef.

Similar Indo-Pacific genera: *Iciligorgia.*

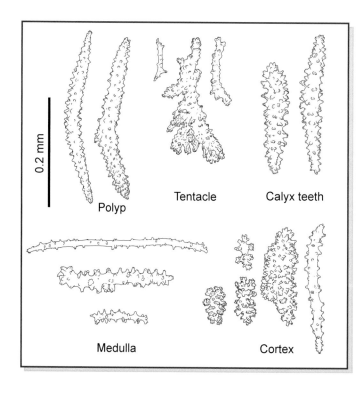

0.2 mm

Polyp   Tentacle   Calyx teeth

Medulla   Cortex

Photos:

1 AND 2: Bali, Indonesia, ca 25 cm tall colonies. *Photos: Roger Steene*

3: Palau, 35 m depth, view ca 10 cm. *Photo: Coral Reef Research Foundation*

4: Close-up of the characteristic, gutter-like branches. Southern GBR, ca 5 cm. *Photo: Gordon LaPraik*

5: Whitsundays, GBR, view ca 8 cm. *Photo: KF*

6: Dried and eroded *Solenocaulon* skeletons are sometimes washed ashore on muddy coasts. Central GBR, length of this fragment ca 20 cm. *Photo: KF*

Alertigorgia

Kükenthal, 1908

Colony shape: Two species with two distinct growth forms are known. The first, *Alertigorgia orientalis* (PHOTOS 1 - 4) is a sparsely branched, bushy gorgonian where the branches are commonly somewhat flattened, with blade-like edges (however, colonies with cylindrical branches are not rare). In many, if not all cases, colonies appear to have begun life as a planula that has settled on a species of the sponge genus *Oceanapia*. The colony forms a thin membrane overgrowing the sponge, and then produces branches above. Large *Alertigorgia* colonies can be found with the basal sponge still alive and apparently unaffected by the association. The second, a new species whose name is to be published shortly, has a growth form that is most unusual in that the branches are grossly expanded and leaf-like (PHOTO 5), and could be mistaken for an alga or a sponge. In all cases, colonies are easily broken because the axis is only made of closely packed sclerites.

Polyps: Monomorphic. In colonies that are expanded leaf-like, the polyps are densely scattered over both faces of the leaves. They form small domes on the surface and have a slit-like mouth opening. In those colonies with slender branches, the polyps are generally bilaterally arranged in two very narrow grooves (essentially the aligning of the slit-like mouths), one of which runs along each blade-like edge of a branch. Where a colony has cylindrical branches, the polyps generally occur in small groups within short, narrow grooves, randomly arranged over the surface. Single, isolated polyps have slit-like mouth openings.

Sclerites: The medulla contains small, thin needles with a few prickles, and thick rods with a few large warts. The cortex contains spindles (usually plump) and ovals, most of which are covered with large complex warts. The polyp head contains curved spindles in a collaret and points arrangement, while short rods and curved, flattened spindles may be found in the tentacles. Colourless.

Colour: Pale brown colony and polyps. Zooxanthellate.

Habitat and abundance: Not uncommon in turbid, silty near-shore waters.

Zoogeographic distribution: Indonesia, Papua New Guinea, eastern and northern Australia.

Similar Indo-Pacific genera: The leaf-like species could be mistaken underwater for *Hicksonella expansa*, the only other gorgonian with this growth form known outside of the Caribbean.

Photos:

1: Northern Great Barrier Reef, ca 30 cm. *Photo: KF*

2: Southern Great Barrier Reef, ca 100 cm. *Photo: Gordon LaPraik*

3: Northern Australia, ca 15 cm. *Photo: Karen Gowlett-Holmes*

4: Papua New Guinea. *Photo: Gustav Paulay*

5: Sulawesi, Indonesia. These flattened forms can be 100 cm tall. *Photo: Leen van Ofwegen*

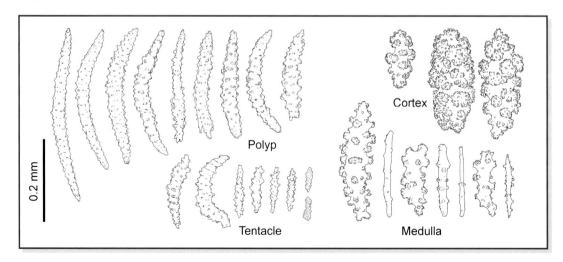

Polyp

Cortex

0.2 mm

Tentacle

Medulla

Erythropodium

Kölliker, 1865

Colony shape: Small polyps arising from a thin, encrusting, basal membrane or a network of ribbon-like stolons. The gastric cavities of the polyps are situated in the upper layer of the membrane, which is separated from the basement layer by a network of large canals. Small, slightly projecting polyp mounds may be present.

Polyps: The polyp bodies are thin, short or long, and the tentacles are very slender.

Sclerites: Small sclerites, all derivatives of 6-radiate capstans, are abundant in the membranes or stolons. Those in the basement layer may be fused into small clumps. The sclerites may be colourless in the top layer and magenta in the basement layer, all colourless, or all magenta.

Colour: Purplish-brown membrane, and brown polyps (*Erythropodium caribaeorum*, Caribbean Sea), pale brown membrane with brown polyps (*Erythropodium hicksoni*, southern Australia), or magenta ribbon-like stolons with grey tentacles (undescribed, Indonesia). Possibly zooxanthellate.

Habitat and abundance: No data are available for the tropical Indo-Pacific form.

Zoogeographic distribution: The only tropical Indo-Pacific species known is as yet undescribed, and was collected in Bali and sent to us by Julian Sprung. The genus is also found in the Caribbean Sea, and in temperate waters of the Indo-Pacific.

Similar Indo-Pacific genera: Stolonic growth forms are also found in *Briareum* and *Rhytisma*.

Photos:

1: The magenta stolons of an *Erythropodium* colony from Bali, Indonesia, growing in a sea water aquarium. *Photo: Julian Sprung*

2: *Erythropodium*, ca 6 cm. *Photo: Julian Sprung*

3: The underside of *Erythropodium*, growing on aquarium glass, ca 10 cm. *Photo: Julian Sprung*

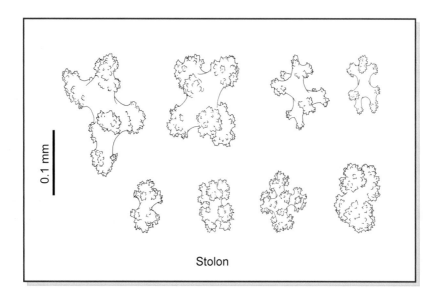

0.1 mm

Stolon

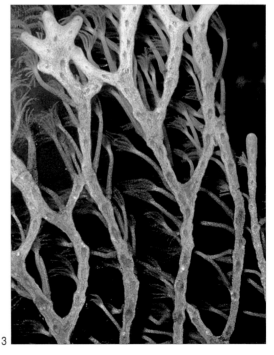

Subergorgia

Gray, 1857

Colony shape: Colonies often quite large, growing in one plane, and laterally to dichotomously branched but not forming nets. All species have long, smooth sclerites, partially fused, embedded in the horny axial medulla. In some species at least, there is a narrow furrow down opposite faces of the branches. Colonies growing in very turbid areas are often covered with strings of mud and detrital material.

Polyps: Monomorphic, medium in size, generally arranged down only two sides of the branches, and retractile into dome-shaped mounds.

Sclerites: Those embedded in the horny axial medulla are narrow, elongate, often branching and fusing together. In the outer cortex the sclerites occur as warty spindles or ovals. The polyps have flattened spindles arranged in 8 points. Brown.

Colour: Light brown to dark reddish-brown, with white polyps. Azooxanthellate.

Habitat and abundance: Relatively common in turbid near-shore environments below 5 m depth, but is also recorded from clearer waters.

Zoogeographic distribution: Most records are from the central Indo-Pacific, but also reported from Madagascar, Mauritius, Zanzibar, Sri Lanka, Japan, Great Barrier Reef and the Solomon Islands.

Similar Indo-Pacific genera: Some pinnate forms may resemble *Pinnigorgia*, however the structure of the axis serves to easily distinguish both taxa.

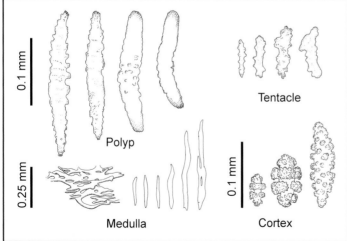

Photos:

ABOVE: Dry colony showing the central furrow and polyps aligned on the branch edges. Width of this colony ca 20 cm. *Photo: KF*

1: Colony growing on the Yongala wreck, central GBR, ca 15 cm. *Photo: KF*

2 AND 3: Branches of *S. suberosa* growing in muddy near-shore environments of the GBR are often covered with detrital material. *Photos: KF*

4: Some large colonies develop irregular growth forms. GBR, ca 100 cm. *Photo: KF*

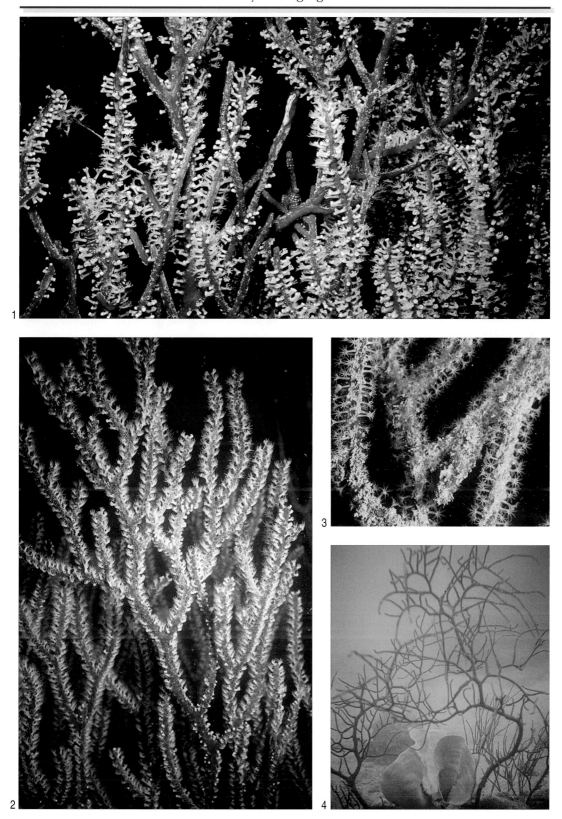

Annella

Gray, 1858

Colony shape: Densely reticulate fans which may be over 2 m tall. The axis consists of horny material, with long, smooth, partially fused sclerites embedded in it. There are only 3 nominal species. In *Annella mollis*, the meshes are generally elongate. In *Annella reticulata*, they are finer and more polygonal, and the retracted polyps appear as low domes. *Annella ornata* is said to be the same species as the earlier described *Annella reticulata*.

Polyps: Monomorphic, retractile, and small, and arranged on all sides of the branchlets.

Sclerites: Those embedded in the horny axial medulla are narrow, elongate, often branching and fusing together. In the outer cortex the sclerites occur as warty spindles, along with numerous, characteristic, small double wheels or double heads. Polyps have slightly flattened spindles arranged as collaret and points. Colourless.

Colour: Sometimes red, but more commonly yellow or pinkish yellow, appearing orange or brown if the polyps are expanded. Azooxanthellate.

Habitat and abundance: On the Great Barrier Reef it is relatively common on current-swept flanks and ridges below the reach of storm waves, all across the shelf. Absent in wave-exposed habitats and areas of low currents.

Zoogeographic distribution: The genus appears to be quite common throughout the coral reefs of the Indian and western Pacific Oceans, and is also found in the Red Sea and Madagascar.

Similar Indo-Pacific genera: Could be mistaken for some reticulate species of the family Melithaeidae that have inconspicuous axial nodes.

Remarks: Until recently these net-like forms were known as species of *Subergorgia*, from which they clearly differ in both sclerite and colony form (Grasshoff 1999).

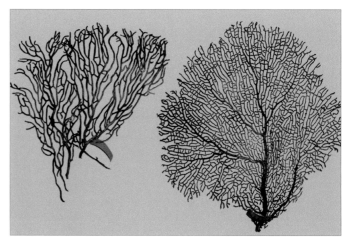

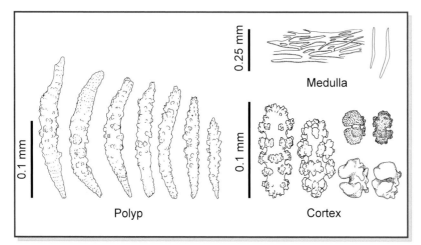

0.25 mm

Medulla

0.1 mm

Polyp

0.1 mm

Cortex

Photos:

ABOVE: *A. mollis* (left) and *A. reticulata* (right). GBR, 20 and 25 cm high. *Photo: PA*

1 AND 2: Spectacular large fans of *Annella* are features of many famous dive sites. *Photos: Roger Steene*

3: Northern GBR, ca 15 cm. *Photo: KF*

4: *Annella* with the typical meandering main axis. Great Barrier Reef, 20 cm. *Photo: KF*

5: Close-up of a colony from Sinai, Red Sea. *Photo: Mark Wunsch*

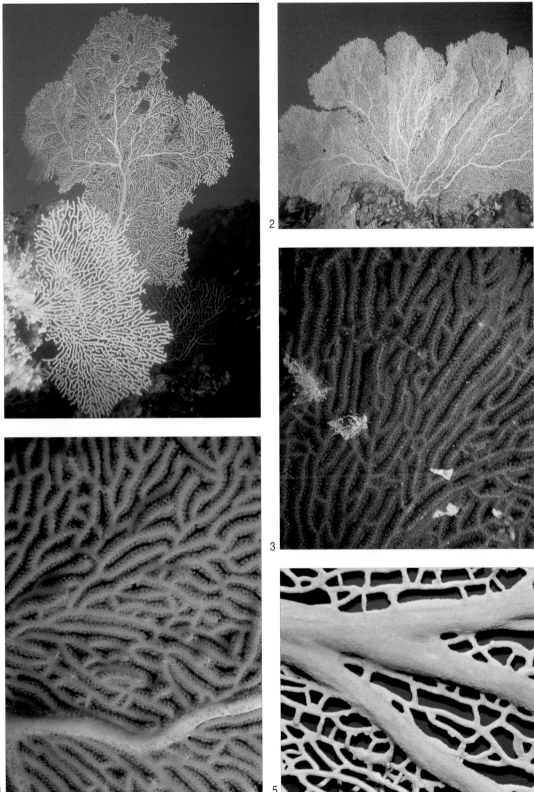

Melithaea, Mopsella, Wrightella, Clathraria, and Acabaria

To be practical, these five genera need to be discussed together. All possess a segmented axis, with conspicuously swollen nodes and straight, rigid internodes, containing cigar-shaped sclerites. With some exceptions, they share the same general appearance, and the same colonial growth forms (planar, bushy, tangled, and often dichotomously branched). Each genus was originally established because of characteristic types of sclerites in the cortex. Since those early days, however, researchers have been able to examine many specimens from numerous locations, which included many colonies with intermediate sclerite shapes, making it difficult or often impossible to decide which genus they belong to. The subsequent generic assignment of such species has widened the original definitions of these five genera to such an extent that there is now considerable overlap between them. It seems likely that all five of the nominal genera will be found to represent the variations of a single genus. If ongoing research will prove this to be the case, then species of *Mopsella, Clathraria, Acabaria* and *Wrightella* will all become species of the first established genus, *Melithaea.*

Polyps: Monomorphic, small, and retractile. Both low and tall calyces may be present in all of the nominal genera: tall calyces are more often seen in *Acabaria*, and a couple of species of *Clathraria* and *Melithaea* are recorded with no calyces at all.

Sclerites: In all species, the sclerites of the axis are short, smooth, cigar-like rods (those in the internodes with a raised middle section), the polyps contain spindle-like and club-like forms arranged as collaret and points, and there are curved, wing-like scales in the tentacles. The cortex contains variations of sclerite form based on the capstan, the club and the spindle. With the exception of some species of *Clathraria* and *Acabaria*, all forms are present in all species, but the amount varies. Clubs occur in the walls of calyces, and are therefore absent in those few species without calyces.

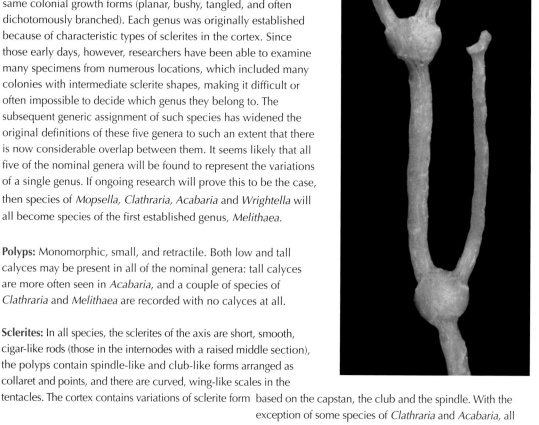

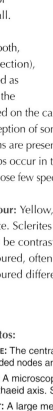

Colour: Yellow, orange, red, dark purple, pink, and white. Sclerites of the calyces and the branch surface may be contrastingly coloured. Axes are usually coloured, often red, and the axial internodes may be coloured differently from the nodes. Azooxanthellate.

Photos:

ABOVE: The central axis of a Melithaeidae, segmented into rounded nodes and straight internodes. Scale: ca 2 cm. *Photo: PA*

LEFT: A microscopic picture of the sclerites in the node of a melithaeid axis. Scale: ca 1 mm. *Photo: PA*

RIGHT: A large melithaeid fan from the GBR. *Photo: KF*

Habitat and abundance: Relatively common under overhangs, in cracks, on steep walls, between bommies, and on current-swept flanks and ridges.

Zoogeographic distribution: Widespread and could reasonably be expected to be found in most areas covered by this book, although only *Acabaria* and *Clathraria* seem to occur in the northern Red Sea. Many species are also found in temperate seas, but the occurrence and diversity is greater in warmer waters.

Similar Indo-Pacific genera: Although there are other genera that have a segmented axis, none have conspicuously swollen nodes. In a few melithaeid species the nodes are not very obvious, and some confusion could arise with species of a similar colony form in other families such as Isididae and Parisididae.

On the following pages, we report the characters of each of the nominal genera, presenting drawings of their most characteristic sclerites, and refer to some examples of overlap.

Melithaea

Milne Edwards & Haime, 1857

Colony shape: Colonies densely branched in one plane forming large fans or bushes of multiple parallel fans. Branchlets within a fan may sometimes fuse, net-like.

Sclerites: The typical cortical sclerites of *Melithaea* are capstans more or less modified as birotulates (double-discs, double-buns), together with knobbed clubs. The longest clubs are found in the calyx walls. The capstans can be simple and quite smooth, or modified with lumps or leafy processes on one side. They can also develop extra elements becoming multirotulate.

A couple of species assigned to this genus have abundant, relatively simple, smooth capstans or birotulates, such as found in *Clathraria rubrinoides*, a bushy colony.

Photos:

ABOVE RIGHT: Central Great Barrier Reef (GBR), ca 6 cm. *Photo: KF*

1: Colourful *Melithaea* from the Philippines, ca 15 cm. *Photo: Doug Fenner*

2: Whitsundays, central GBR, ca 20 cm. *Photo: KF*

3: Cluster of large colonies, southern Papua New Guinea. *Photo: KF*

4: *Melithaea* serving as perches for crinoids. Central GBR, ca 80 cm. *Photo: KF*

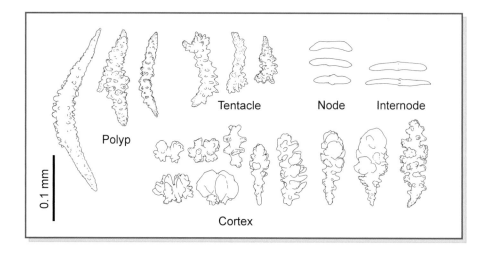

Tentacle Node Internode

Polyp

0.1 mm

Cortex

Mopsella

Gray, 1857

Colony shape: Colonies densely branched in one plane, forming large fans or bushes of multiple parallel fans. Branchlets within a fan may sometimes fuse to form nets.

Sclerites: The most characteristic sclerites are leaf-clubs and leaf-spheroids. The leafy processes may be thick and blunt, or thin and sharp like broad blades. The longest clubs are found in the calyx walls.

Some species placed in this genus have mostly leafy spheroids with short roots, and very few long handled clubs, such as found in some species assigned to *Clathraria*.

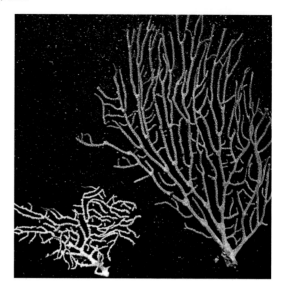

Photos:

ABOVE, AND 4: Two species from the Great Barrier Reef (GBR), preserved and in the field, ca 12 cm. *Photo: KF*

1: One of the many colour forms of *Mopsella*. GBR, ca 30 cm. *Photo: Queensland Museum*

2: Southern GBR, ca 10 cm. *Photo: Gordon LaPraik*

3: Whitsundays, GBR, ca 8 cm. *Photo: KF*

5: Large colony attached to a steep rock face. *Photo: Queensland Museum*

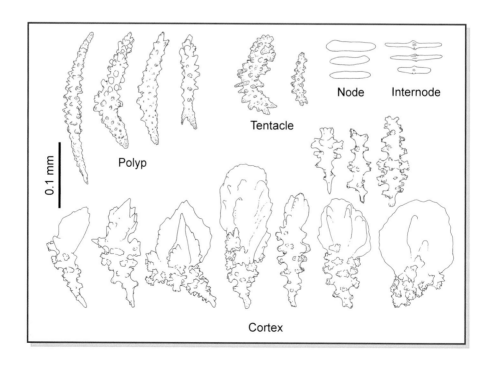

Node   Internode

Tentacle

Polyp

0.1 mm

Cortex

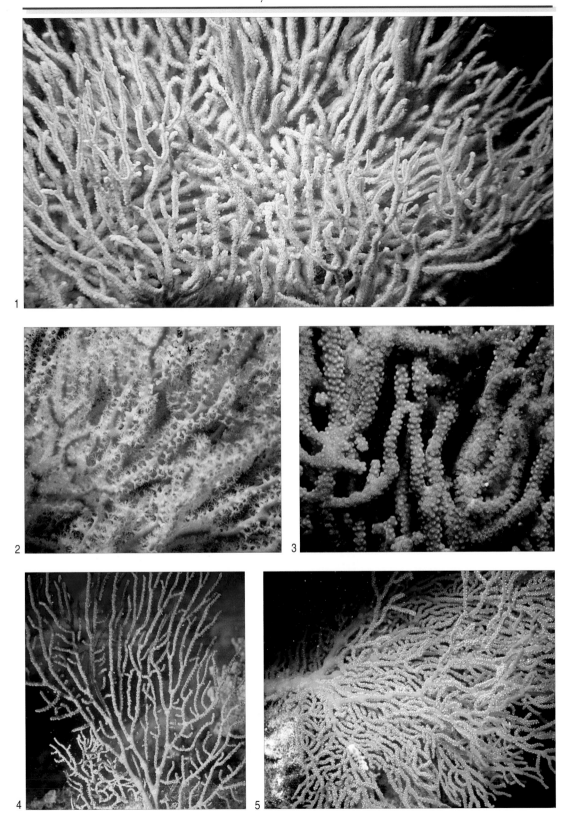

Wrightella
Gray, 1870

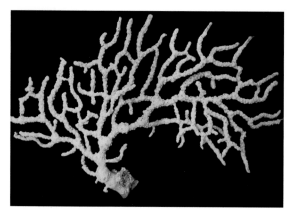

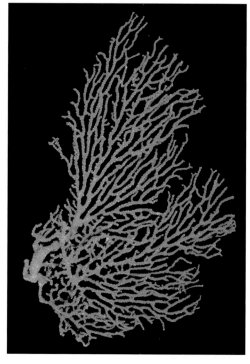

Colony shape: Small fans. Branches may join to form nets. Very few specimens recorded.

Sclerites: The most characteristic sclerites are leafy spheroids, a number of which are very large. The leaf-like processes are usually very short and can be ridge-like. The longest clubs are found in the calyx walls.

Smaller forms of leafy spheroids are like those found in some species assigned to *Clathraria*, and those with a handle-like root are similar to the leaf-clubs in *Mopsella*.

Photo:

ABOVE AND LEFT: Preserved colonies of *Wrightella*. *Photos: PA*

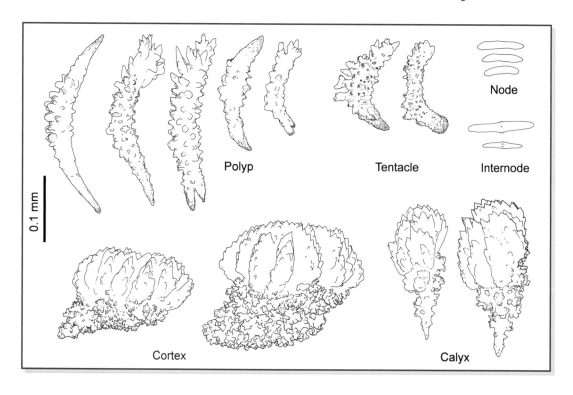

Gray, 1859

Colony shape: *Clathraria rubrinodis*, the first named species of this genus, repeatedly branches in a dichotomous manner in all directions, producing a very bushy colony (see Photo Right). The surface of the branches is flat, with no calyces. Other species that form upright fans, or groups of fans, have since been added to the genus.

Sclerites: The characteristic sclerites of *C. rubrinodis* are knobby capstans, some slightly modified on one side.

In other species assigned to the genus, there are low calyces with leaf-clubs in their walls, and in others the capstans are conspicuously modified on one side approaching the leafy spheroids of *Mopsella*. There are also fan-shaped species of *Melithaea* with abundant capstans.

Photo:

Right: *Clathraria rubrinodis* from the Red Sea. *Photo: Manfred Grasshoff*

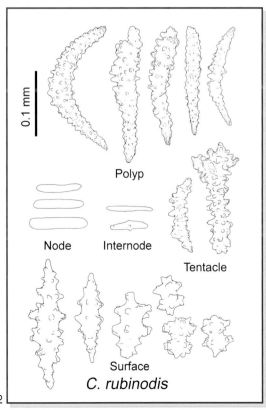

0.1 mm

Polyp

Node   Internode

Tentacle

Surface

C. rubinodis

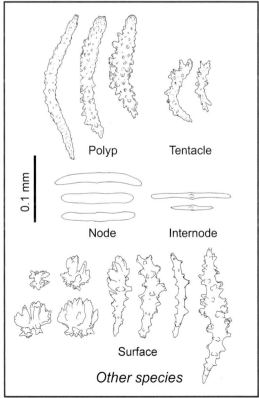

0.1 mm

Polyp   Tentacle

Node   Internode

Surface

Other species

2

Acabaria

Gray, 1859

Colony shape: *Acabaria divaricata*, the first named species of the genus, branches dichotomously in one plane. Other species that form bushes or untidy tangles of branches have been added to the genus. Calyces are usually present, and sometimes very tall.

Sclerites: The characteristic and dominant sclerites are spindles. These spindles are often developed along one side with lump-like or leafy-like processes. Leaf clubs occur in the calyx walls.

In some species assigned to the genus, the shorter sclerite forms are oval, and often dominant, and resemble the leafy spheroids found in *Mopsella* and *Clathraria*.

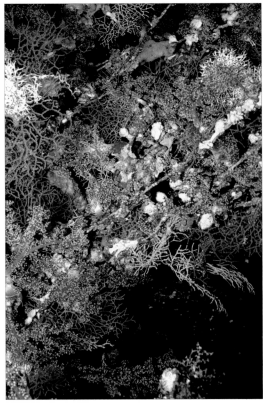

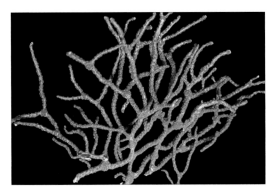

Photos:

ABOVE: Growth on the pillars of a jetty, dominated by *Acabaria* and *Dendronephthya*. Gulf of Aqaba, northern Red Sea. *Photo: KF*

ABOVE LEFT: Stout colony from the GBR, preserved, ca 13 cm. *Photo: KF*

1: Tangled growth form, Gulf of Aqaba, Red Sea. *Photo: KF*

2: Fan-shaped colony, northern Australia. *Photo: PA*

3: Stout branches, Saudi Arabia, Red Sea, ca 20 cm. *Photo: Emre Turak*

4: Southern GBR, ca 20 cm. *Photo: Gordon LaPraik*

5: Juvenile colony, northern Red Sea, ca 3 cm. *Photo: Mark Wunsch*

6: Coloured nodes are characteristic for some *Acabaria*. Northern Red Sea. *Photo: Günther Bludszuweit*

0.1 mm

Node

Tentacle

Internode

Polyp

Cortex

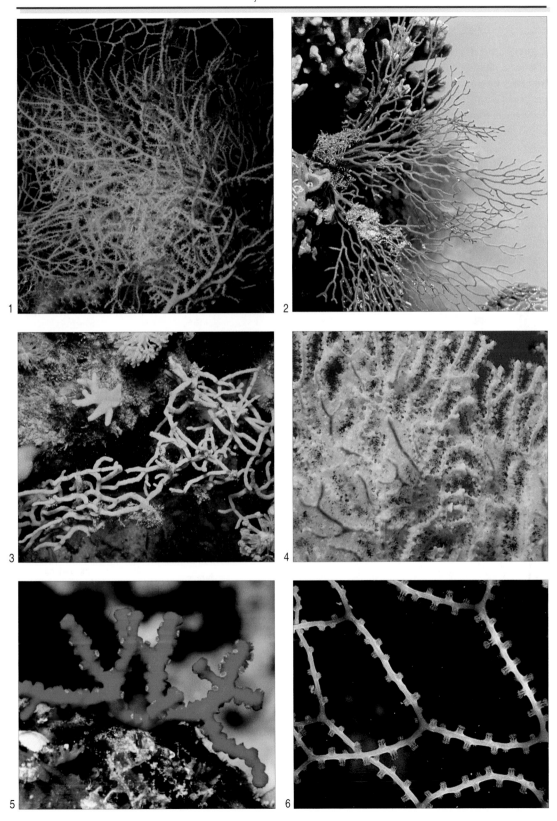

Parisis

Verrill, 1864

Colony shape: Profusely branched fans growing in one plane. In some species the branchlets are markedly thinner than the stem and main branches. The segmented axis is composed of solid, longitudinally grooved, calcareous internodes formed from fused tuberculate sclerites, which alternate with horny nodes containing sclerites in the form of lobate rods. Branching emanates from the calcareous internodes. In some species the dark nodes can be seen through the surface of the thicker branches. The polyps are distinct, contracting and withdrawing the tentacles to form volcano-shaped mounds, and occur all around the branches. Colonies feel hard, and generally will not bend very far before they snap. Some species are encrusted with sponges.

Polyps: Monomorphic, rigid, and non-retractile.

Sclerites: The polyp wall and the colony surface are covered with tuberculate, plate-like sclerites, sometimes mixed with smaller irregular-shaped forms. There are no tentacle sclerites. Under a low-power microscope, the sclerites in most species can generally be seen fitting close together forming a pavement. The sclerites of the soft axial nodes are lobate rods. Colourless.

Colour: Pink, yellow, pale blue. Azooxanthellate.

Habitat and abundance: Rare in shallow water.

Zoogeographic distribution: Generally throughout the broader Indo-Pacific region in deep water. Shallow-water taxa recorded only from the Red Sea, the Great Barrier Reef and New Caledonia.

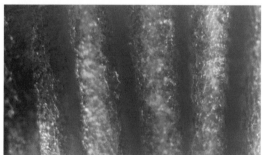

Similar Indo-Pacific genera: The jointed axis could cause confusion with genera from the families Isididae and Melithaeidae. Plate-like sclerites are also found in *Keroeides* and *Paracis*, but neither of these have a segmented axis.

Photos:

TOP: Segmented axis of stem (left; length: 30 mm), and branch (right, 15 mm). *Photos: PA*

ABOVE, AND 3: Close-up of the vertical ridges of an axial node, and the embedded sclerites. *Photos: PA*

1 AND 2: Dredged ca 40 cm wide colony from the GBR, and close-up of branches. *Photos: Marine Bioproducts Group, AIMS*

4 AND 6: New Caledonia. Scale bar: 5 cm. *Photos: Manfred Grasshoff*

5: Mindanao, Philippines, 35 m depth. *Photo: Coral Reef Research Foundation*

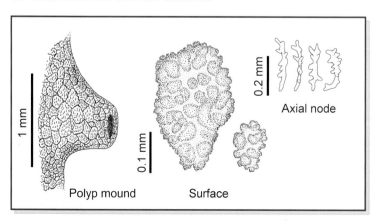

Axial node

1 mm

0.1 mm

0.2 mm

Polyp mound   Surface

Keroeides

Studer (and Wright), 1887

Colony shape: Richly branched, delicate fans, with erect, conical polyp calyces that are generally not crowded. The axis has outer layers of smooth, spindle-shaped sclerites.

Polyps: Retractile into prominent calyces.

Sclerites: On the polyp heads, there are many small to large tuberculate rods that are arranged obliquely in 8 double rows or points, without a collaret. In the tentacles there are numerous small, branched rods or scales. The sclerites of the surface of the stem and branches are large spindles, often flattened, and irregularly shaped plates. The outer layer of the axis is tough, and consists of smooth spindle-shaped sclerites tightly bound in a dense gorgonin matrix. Underneath this, surrounding the chambered central core, the sclerites form a thick layer with very little gorgonin present. The sclerites of the calyx wall are usually the same form as those of the branch surface, but they may be remarkably smaller, in which case they are generally arranged longitudinally. Colour red, orange or colourless. In the outer layer these sclerites are tightly bound in a gorgonin matrix.

Colour: Red, orange, or white. Azooxanthellate.

Habitat and abundance: Rare in shallow water where it has been found at the more extreme diveable depths on steep walls and overhanging rock.

Zoogeographic distribution: Collected, mostly by remote equipment, from the Red Sea, Mayotte, the Moluccas, Japan, Palau, New Guinea, Funafuti, Tuvalu, northern Australia, New Caledonia, Northern Mariana Is. and Hawaii.

Similar Indo-Pacific genera: *Parisis*, *Paracis*, and several genera in the family Plexauridae in which the surface sclerites are large and form a pavement-like layer.

Photo:
1: *Keroeides gracilis*, New Caledonia. The scale bar represents 5 cm. *Photo: Manfred Grasshoff*

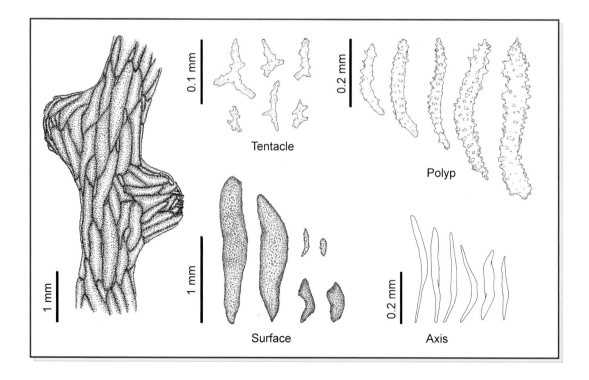

0.1 mm

Tentacle

0.2 mm

Polyp

1 mm

1 mm

Surface

0.2 mm

Axis

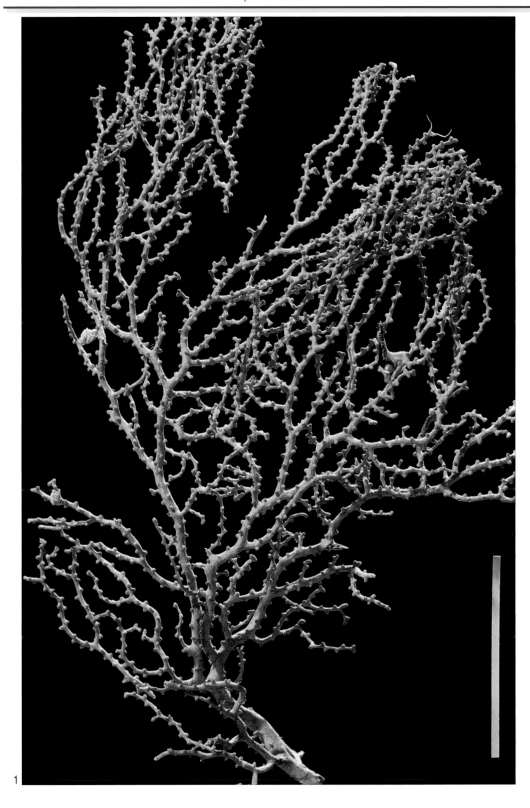

1

Acanthogorgia

Gray, 1857

Colony shape: Colonies are commonly planar and reticulate and up to about two thirds of a meter high. Others form dense, bushy tangles, or loose, laterally branched fans. Colonies are generally delicate with thin branches. The coenenchyme layer between the polyps is usually so thin that the black axis can be seen through it.

Polyps: Monomorphic, non-retractile, often very tall when contracted, and sometimes trumpet-shaped. Such contracted structures look like calyces, but they are actually the non-retractile polyps.

Sclerites: The polyps are covered with warty spindles, commonly boomerang-shaped, and usually arranged along the body wall in 8 double rows. If the tip of a spindle protrudes, it is smooth. In many species, the spindles at the top of the polyp, just below the tentacle bases, are very long and project to form a conspicuous spiny "crown". The backs of the tentacles are densely covered in small, flat, boomerang-shaped sclerites. The thin surface tissue (coenenchyme) contains small, warty spindles, which may occur with thornstars or capstan derivatives. Sclerites are always colourless.

Colour: Bright colours such as yellow, orange, pink, red, or dark purple. Colonies are often multicoloured, and the colour of the polyps may contrast to that of the coenenchyme. Azooxanthellate.

Habitat and abundance: Uncommon. Found on deeper slopes, steep walls, and current-swept flanks and ridges.

Zoogeographic distribution: This genus appears to occur in most areas covered by this book. It is also reported from deep waters in most of the World's oceans.

Similar Indo-Pacific genera: Some genera of the family Plexauridae, such as *Astrogorgia*, in which the polyp calyces could be mistaken for polyps.

Remarks: The sclerite architecture in this group is remarkably uniform. Nominal species with different growth form and colour, from different parts of the world, may have sclerites that show little difference in size and shape. Many species of *Acanthogorgia*, in which the polyp sclerites are so short that there is little or no projecting crown, erroneously appear in the literature under the name *Acalycigorgia* Kükenthal, 1919. *Acalycigorgia* is actually a synonym of *Acanthogorgia*, and most species referred to the former should be called *Acanthogorgia*.

Photos:

1: Cave on a reef in Sinai, Egypt. *Photo: KF*

2, 6 AND **7:** GBR, 3 - 10 cm. *Photos: KF*

3: Preserved polyp with crown, 2.3 mm. *Photo: PA*

4: Sabah, Malaysia, ca 15 cm. *Photo: Frances Dipper*

5: Dried colony from the GBR, showing the axis through the dried tissue; ca 5 cm. *Photo: KF*

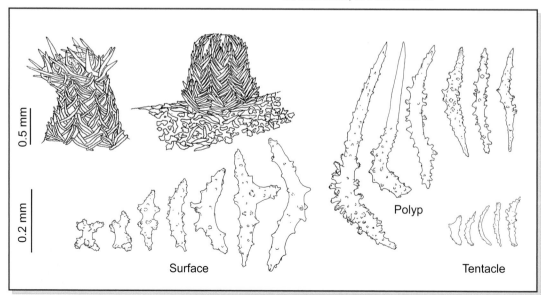

0.5 mm

0.2 mm

Surface

Polyp

Tentacle

Anthogorgia
Verrill, 1868

Colony shape: Generally fan-like, up to about 1 meter tall, and commonly with free branches and few net-like branch fusions. However, some slightly bushy colonies have been recorded, and large fans in strong currents may develop many net-like joins, particularly involving the major branches. The coenenchyme layer between the polyps is thick and obscures the axis.

Polyps: Monomorphic, non-retractile, cylindrical or dome-like, and often tall relative to *Muricella* (page over). As with *Acanthogorgia* (previous page) the contracted polyps can be mistaken for calyces.

Sclerites: The polyps are covered with blunt spindles, with large warts, that are arranged in angled double rows (chevrons), often somewhat irregularly, along the body wall. The tentacles contain small rods. The surface tissue (coenenchyme) contains spindles similar

to those in the polyps, with ovals, small capstans and capstan derivatives. Sclerites may be coloured.

Colour: Red, brown, blue, pink, and yellow, sometimes with coloured sclerites. Polyp colour may contrast to the coenenchyme colour. Azooxanthellate.

Habitat and abundance: Relatively rare, occurring on steep slopes, channels, and canyon walls.

Zoogeographic distribution: Not many species recorded, but probably found in most areas covered by this book.

Similar Indo-Pacific genera: Could be confused with some genera of the family Plexauridae, such as *Astrogorgia*, where the polyp calyces could be mistaken for polyps. This genus could be the same as *Muricella* (page over). Dry colonies may resemble some species of *Nicella*.

Photos:
LEFT: Dried colony, northern Australia, ca 15 cm. *Photo: KF*
1: Colony from Sulawesi, Indonesia. *Photo: Leen van Ofwegen*
2 AND 3: Preserved colonies, showing the typical growth form of free-ending branches. *Photos: PA*
4: Philippines. *Photo: Leen van Ofwegen*
5: Hong Kong, 5 cm. *Photos: KF*
6: Sulawesi, Indonesia, ca 5 cm. *Photo: Roger Steene*
7: Preserved *Anthogorgia* polyp, ca 2 mm. *Photo: PA*

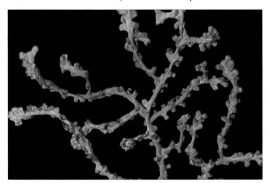

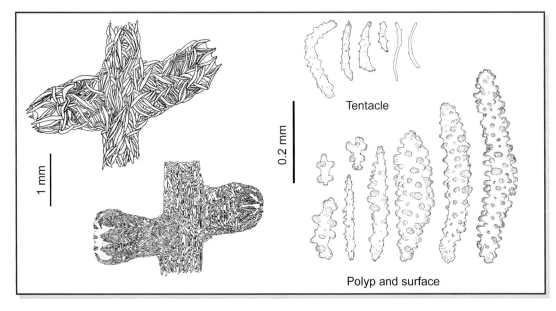

1 mm

0.2 mm

Tentacle

Polyp and surface

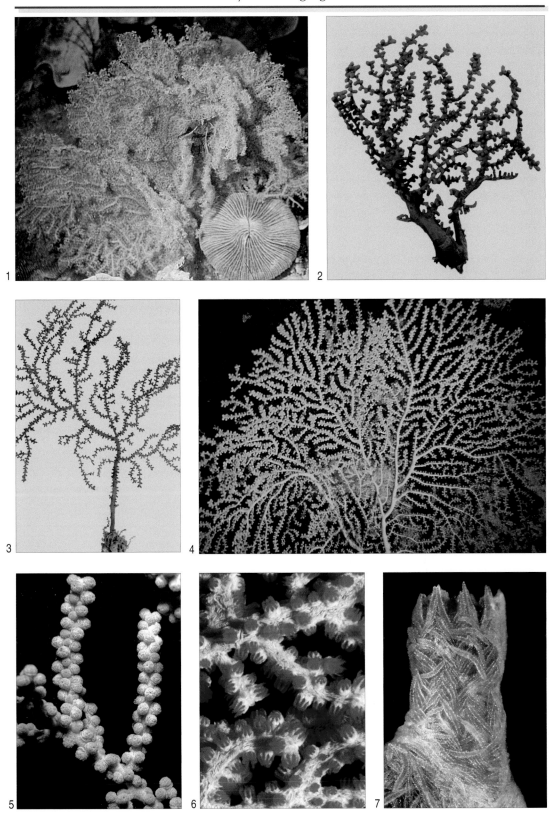

Muricella

Verrill, 1869

Colony shape: Planar fans, often net-like, up to at least two-thirds of a meter tall. The axis of the thicker branches tends to be flattened at right angles to the plane of the fan, and in old colonies small branches often bend and grow out perpendicular to the fan. The coenenchyme layer between the polyps is thick and obscures the axis.

Polyps: Relative to *Anthogorgia*, the polyps are short and dome-shaped. They are monomorphic and non-retractile. As with *Acanthogorgia* and *Anthogorgia*, the contracted polyps can be mistaken for calyces.

Sclerites: The polyp tentacles contain small rods, and the polyp body is covered with blunt spindles with large warts. These spindles tend to be arranged along the body wall in angled double rows (chevrons), but in preserved or dried specimens this arrangement can often only be seen at the polyp summit, below the tentacle bases. The surface tissue (coenenchyme) contains small capstans, and spindles similar to those in the polyps, but many of these spindles are very large, commonly larger than the width of the branch they are on. Sclerites may be coloured.

Colour: Brown, yellow, pink and white. Polyp colour may contrast to the coenenchyme colour. Azooxanthellate.

Habitat and abundance: Rare, occurring on slopes, channels, and canyon walls.

Zoogeographic distribution: Not many species recorded, but could reasonably be expected to be found in most areas covered by this book.

Similar Indo-Pacific genera: *Anthogorgia,* and some genera of the family Plexauridae, such as *Astrogorgia,* in which the calyces could be mistaken for polyps.

Remarks: Not much separates this genus from *Anthogorgia,* and a study of a large suite of specimens could see the two synonymised. The characters of the genus *Muricella* were misunderstood for many years, and most species with this name in the literature have polyps that retract into well-formed calyces, and should really be placed in *Astrogorgia*.

Photos:

1: A more than 1 m large fan, with unusual net-like branch fusions. Whitsundays, central GBR. *Photo: Götz Reinicke*

2: Solomon Islands. *Photo: Gary Williams*

3: Papua New Guinea. *Photo: Coral Reef Research Foundation*

4: Sabah. *Photo: Frances Dipper*

5: Base of the colony depicted in fig. 1, showing the net-like growth form. *Photo: KF*

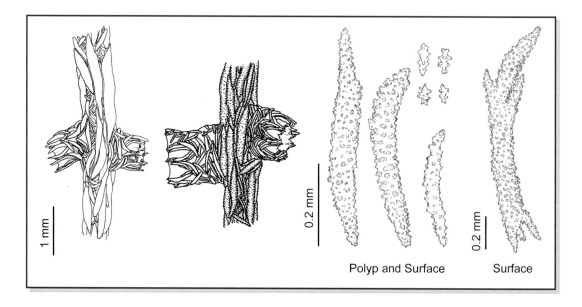

1 mm

0.2 mm

0.2 mm

Polyp and Surface   Surface

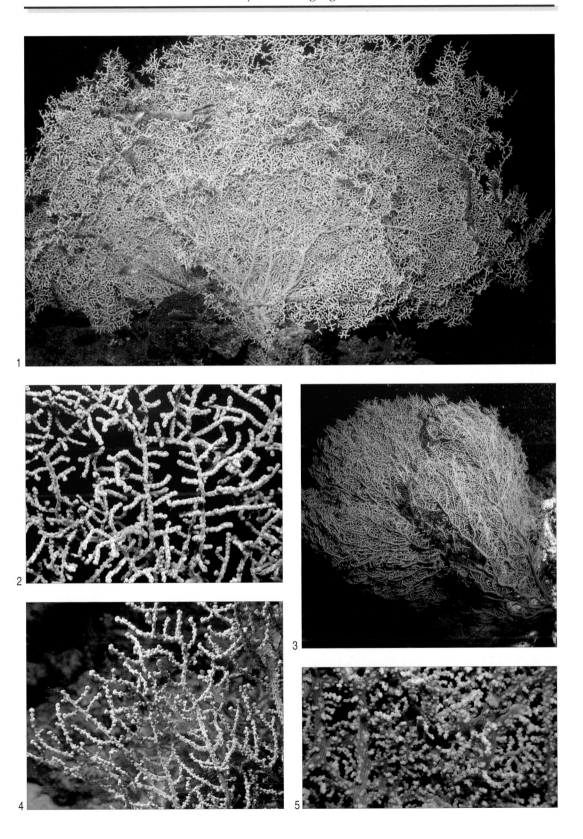

Euplexaura

Verrill, 1869

Colony shape: Colonies mostly grow as fans in one plane, and are generally not richly branched. The stem, main branches and branchlets are usually quite thick, and all of a similar diameter. Branchlets, which are often short, tend to rise more or less at right angles and then bend upwards, and the branch tips are often swollen. Small colonies can look somewhat like an ornate candelabrum. A few species have a more bushy growth form, and may have long whip-like branches.

Polyps: Monomorphic, retractile, and sometimes very large leaving large apertures in the branch surface. Calyces may be present, and quite large, or they may be completely absent.

Sclerites: The polyp head may have large spindles in a strong collaret and points arrangement, a few irregularly placed spindles, or no sclerites at all. The surface (coenenchyme) contains robust oval or sub-spheroidal sclerites, and, sometimes, plump spindles. Surface sclerites are usually densely covered with large complex warts. Colourless.

Colour: Yellow, pink, red, orange, purple, brown, grey, grey-green, or white. Azooxanthellate.

Habitat and abundance: Uncommon but not rare, and grows in both clear and turbid waters.

Zoogeographic distribution: Widespread and may occur in most areas covered by this book.

Similar Indo-Pacific genera: Colonies of *Isis hippuris* can have similar growth forms.

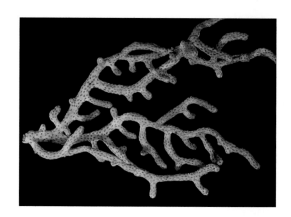

Photos:

Above: Dried specimen from the GBR. *Photo: KF*

1, 3 AND **4:** Hong Kong. *Photos: KF*

2: Colony (ca 50 cm tall) growing on inter-reefal sea floor of the central GBR. *Photo: KF*

5: Yongala Wreck, central GBR. *Photo: KF*

6: Same species as **5**, dried. *Photo: KF*

7: The probably most typical growth form of *Euplexaura*, with thick branches. GBR. *Photo: KF*

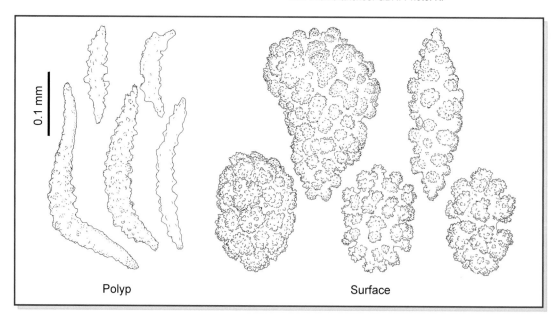

0.1 mm

Polyp   Surface

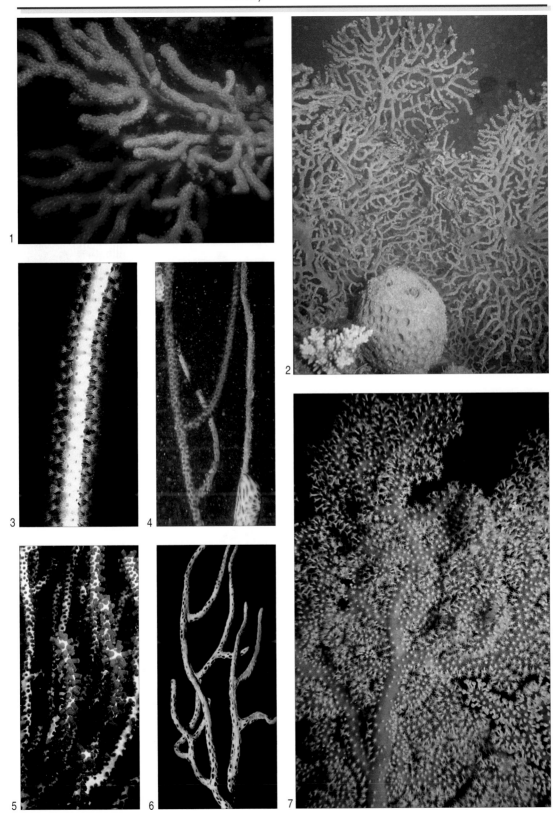

Bebryce

Philippi, 1841

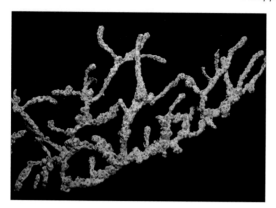

Colony shape: Colonies can grow as fans or bushes, and reach sizes up to at least 50 cm diameter. Branching is lateral, irregular, and often untidy. Most *Bebryce* colonies are overgrown by an encrusting sponge.

Polyps: Monomorphic, and retractile within calyces that may be relatively prominent.

Sclerites: The characteristic sclerites, unique to this genus, are called (among many other names) rosettes. They have a warty base from which projects a clump of thorns, antlers, spines, or bristles. These sclerites occur on the surface of the colony coenenchyme and on the calyces. On the calyces they may become modified as spiny clubs so as to fit better to the contours of the walls. Also in the surface, and

dominating the deeper layers, are 3- to 6-rayed stellate plates, often disc-like, that usually have a central warty boss on one side. The polyp head has large sclerites in a well-formed collaret and points arrangement. The collaret sclerites are curved or bow shaped, some developing a middle arm, and the point sclerites often have a thorny tip. The tentacles contain curved dragon-wing-like scales. Sclerites are always colourless.

Colour: Yellow, white, brown, and red. In many, perhaps the majority of cases, the colour may be due to an overgrowing sponge. Azooxanthellate.

Habitat and abundance: Uncommon in shallow water. Reported from both clear water and turbid environments.

Zoogeographic distribution: Widespread and probably occurs in most areas covered by this book.

Similar Indo-Pacific genera: none.

Remarks: On a microscope slide preparation, rosettes with short spines and a heavy base may sit vertically, rather lie on their sides, making detection difficult. So, the sclerites need to be rolled while being viewed. In some species the rosettes may not be very common and can be overlooked.

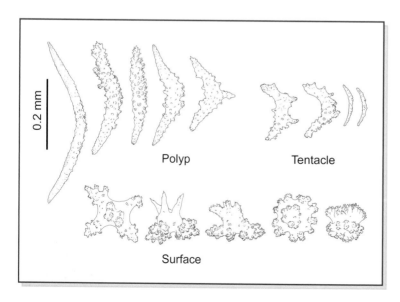

0.2 mm

Polyp

Tentacle

Surface

Photos:

ABOVE: Preserved sample from the northern GBR. *Photo: KF*

1, 3, AND **4**: GBR. *Photos: KF*

2: Preserved colony, GBR, scale bar in centimeters. *Photo: KF*

5: Same colony as in **2**. *Photo: KF*

Echinomuricea

Verrill, 1869

Colony shape: Colonies up to about 50 cm tall, generally with few, long and slender, whip-like branches, and growing in one plane or as loose bushes. Branches often arise at nearly right angles and then curve upward. In a number of species the branches are unusually thick.

Polyps: Monomorphic, and retractile into low or tall calyces.

Sclerites: The characteristic sclerites are the thornscales of the calyces, which have a single, long spine arising from a spreading, warty base. Modified forms of these with a large base, or a couple of spines, can also be found in the calyces and in the colony surface, along with warty spindles and branched forms. The spine on a thornscale can be very smooth, or it may be modified with smaller spines. The polyp head contains large spindles and rods in a collaret and points arrangement. Sclerites are usually red, but they are colourless in some species.

Colour: Usually red, but yellow and white are also reported. Azooxanthellate.

Habitat and abundance: Uncommon. Known to occur in both clear and turbid water.

Zoogeographic distribution: Central Indo-Pacific, and New Caledonia. There are a number of species in the literature reported from farther afield whose true identity needs to be verified.

Similar Indo-Pacific genera: *Trimuricea, Echinogorgia*.

Photos:
1 AND 2: Hong Kong, ca 10 cm. *Photos: KF*
3 AND 4: Northern GBR, 15 - 20 cm. *Photos: KF*
5: Preserved colony from PHOTO 3. *Photo: KF*
6: Preserved colony from PHOTO 4. *Photo: KF*

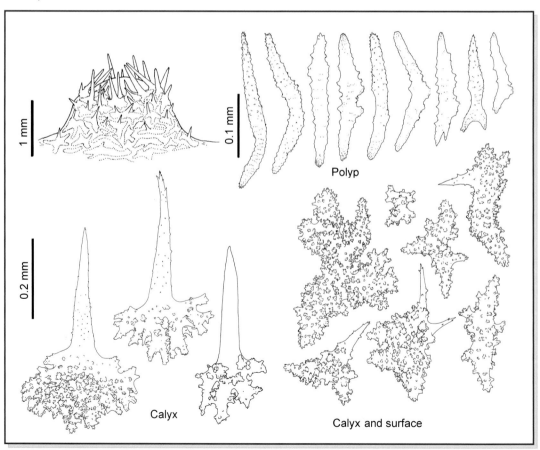

1 mm

0.1 mm

Polyp

0.2 mm

Calyx

Calyx and surface

Trimuricea

Gordon, 1926

Colony shape: Colonies are irregularly, laterally branched in one plane, often with many branch fusions that produce net-like fans up to about 50 cm tall.

Polyps: Monomorphic and retractile, with well-developed calyces that are often very crowded.

Sclerites: The polyps have very unusual sclerites arranged as collaret and points. Many of the sclerites are triradiates, with a horizontal bar forming an element of the collaret, and a vertical prong forming an element of the points. Unbranched, curved sclerites fit in amongst these triradiates. The sclerites of the calyces are thornscales. The base of the scale consists of 2 - 4 warty arms, and a central, relatively smooth, spine, which may have 1 - 2 smaller spines at its base, dominates the upper part. Because the calyces are usually so crowded, there is very little surface tissue (coenenchyme) between them. It contains thorn scales, irregularly branched warty sclerites, and spindles that sometimes have one or more smooth spines. Sclerites are colourless.

Colour: The live colour is recorded for only one species, which is red.

Habitat and abundance: Rare.

Zoogeographic distribution: Mozambique, Andaman Islands, Mergui Archipelago and New Caledonia.

Similar Indo-Pacific genera: *Echinomuricea* and *Echinogorgia*.

Photos:
1: Preserved colony from New Caledonia. *Photo: Manfred Grasshoff*
2: Close-up of **1**. *Photo: Manfred Grasshoff*

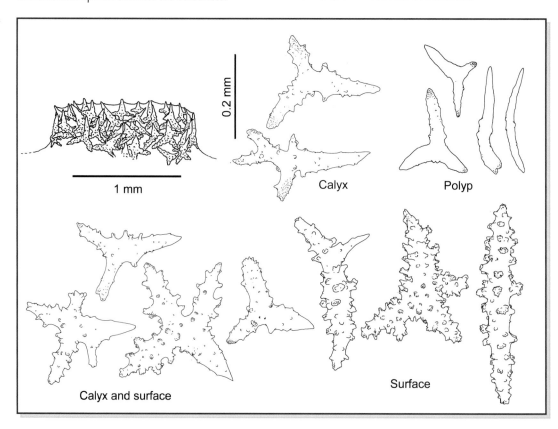

0.2 mm

1 mm

Calyx

Polyp

Calyx and surface

Surface

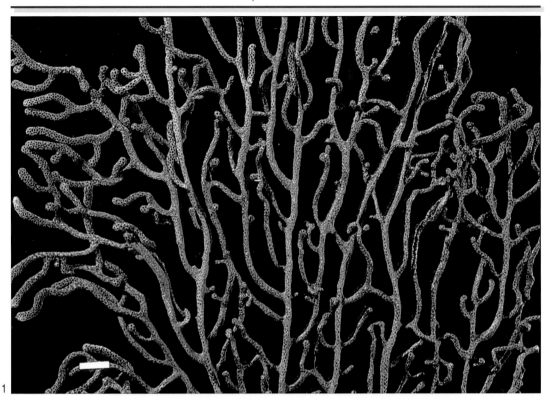

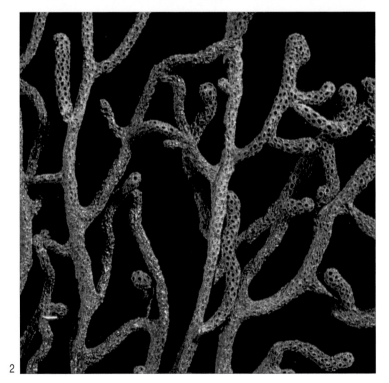

Paracis

Kükenthal, 1919

Colony shape: Richly branched planar fans, branches generally thin.

Polyps: Monomorphic, and retractile into prominent calyces.

Sclerites: The colony surface between the calyces is covered with a pavement-like layer formed by large plates or flattened spindles with a lumpy surface. The calyces contain thornscales, which have a flattened root-like structure from which a blunt projection arises that usually has a denticulate or spiny surface. The polyp head has large sclerites in a collaret and points arrangement. The collaret sclerites are bow-shaped spindles. The points are formed from hockey stick shaped sclerites with a thorny extremity. Thorny clubs can be found in the very base of the tentacles,

and dragon-wing scales in the distal portion. Sclerites can be dark pink to wine-red, though tentacle sclerites may be paler, and colourless. The centre of the large surface plates is often darker than the edges.

Colour: Usually dark pink to red wine coloured, but also white. Azooxanthellate.

Habitat and abundance: Uncommon. Grows in deeper waters, usually below 50 m.

Zoogeographic distribution: Central Indo-Pacific and New Caledonia. Although a number of nominal species are recorded from farther afield, their identity still needs to be verified.

Similar Indo-Pacific genera: *Parisis, Keroeides.*

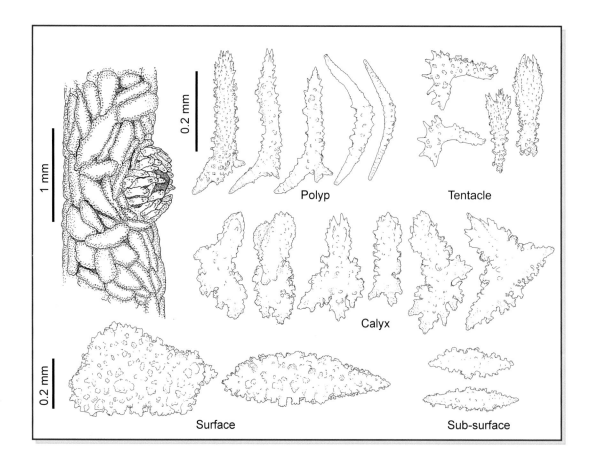

Polyp

Tentacle

Calyx

Surface

Sub-surface

Photos:

1 AND **2:** Sabah. *Photo: Frances Dipper*

3 AND **4:** Preserved colony from New Caledonia. *Photo: Manfred Grasshoff*

Villogorgia

Duchassaing & Michelotti, 1860

Colony shape: Richly branched fans with thin branchlets that occasionally join together to form nets.

Polyps: Monomorphic, and retractile into prominent calyces.

Sclerites: The characteristic sclerites are the calicular thornscales. They have a broad, flat, multi-armed base from which arises a short, globular, spiny or leafy projection. Sclerites with more or less the same style of projection, but modified as thornstars or thornspindles, occur in the coenenchyme. The polyp head has large sclerites in a collaret and points arrangement. The collaret sclerites are bow-shaped spindles, and the points are formed from hockey stick shaped sclerites with a thorny extremity. The tentacles contain curved dragon-wing-like scales. Sclerites are recorded as dark red, amber, pale yellow and colourless.

Colour: Dark red, yellow, brown, or orange-brown. Azooxanthellate.

Habitat and abundance: Uncommon in shallow water, but relatively common at greater depths, and found in both clear and turbid water.

Zoogeographic distribution: Widespread, and probably occurs in most areas covered by this book. Two species are described from the Atlantic.

Similar Indo-Pacific genera: *Echinogorgia, Echinomuricea, Trimuricea, Acanthogorgia.*

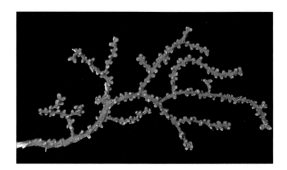

Photos:
ABOVE: Preserved fragment from the GBR, ca 7 cm. *Photo: KF*
1, 2, AND 5: Sabah. *Photos: Frances Dipper*
3: Torres Straits, ca 20 cm. *Photo: KF*
4: GBR; *Photo: Queensland Museum*
6: Central GBR. *Photo: KF*

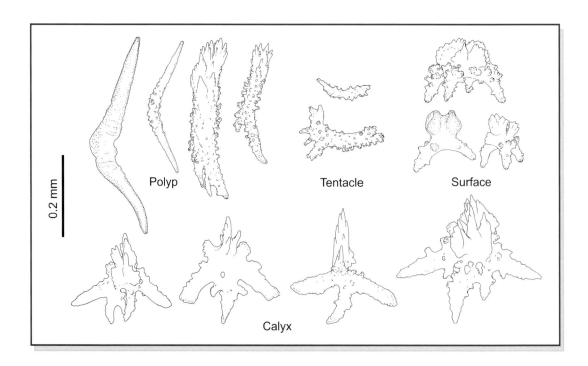

0.2 mm

Polyp Tentacle Surface

Calyx

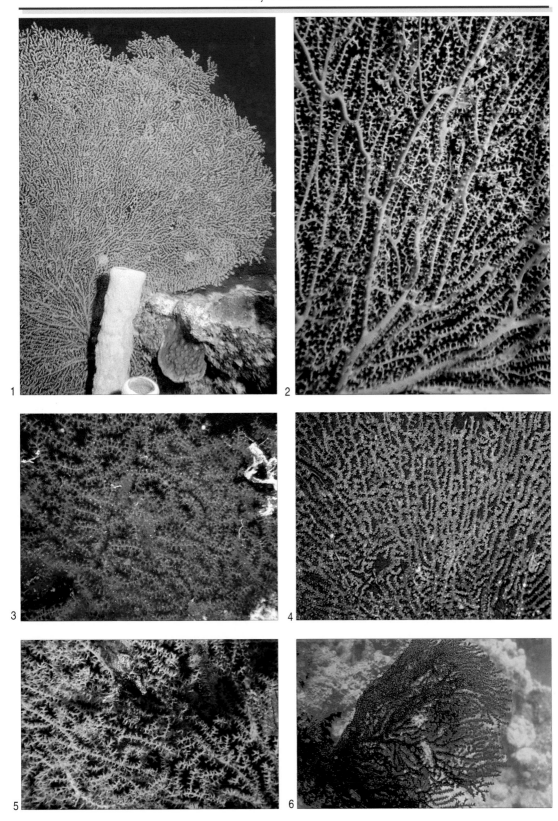

Echinogorgia

Kölliker, 1865

Colony shape: Colonies grow in one plane. The main branches always produce very short side branches, of which at least a few are nearly always fused together. Commonly many branches fuse to produce net-like fans. The branches are usually not very thick, but the colonies can grow quite large.

Polyps: Monomorphic, and completely retractile into spiny calyces that can be very prominent.

Sclerites: The calyces are formed from numerous, rather thick, well-formed thornscales. They generally have about three thorn-like or blade-like projections and a complex tuberculate root structure, but it is not uncommon to find small scales with a single projection, and large ones with many. In some species, the calyces are so numerous and close together that there is barely room for any other type of sclerite on the branch surface. In those species in which the calyces are spread out, the outer layer of the branches contains sclerites that are often called rooted leaves, though the outward pointing projections may be less leaf-like and more spine-like or

(continued)

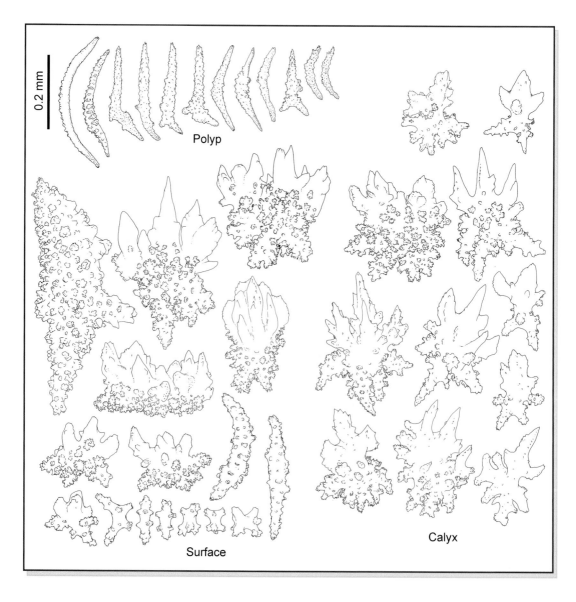

Polyp

Surface

Calyx

0.2 mm

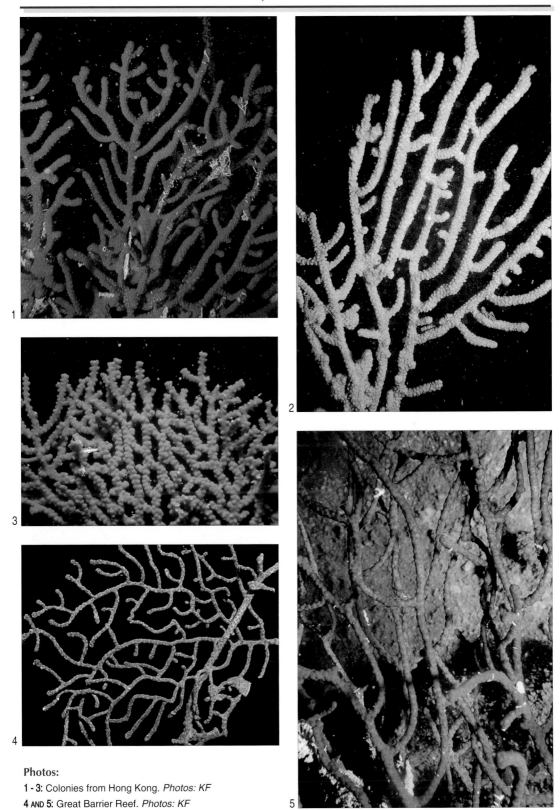

Photos:
1 - 3: Colonies from Hong Kong. *Photos: KF*
4 AND 5: Great Barrier Reef. *Photos: KF*

6

appear in the literature under a number of different generic names, the range of the group is unclear, but it does appear to be widespread. It is known to occur in New Caledonia, tropical and subtropical Australia, Papua New Guinea, some parts of Indonesia, Singapore, Philippines, Sri Lanka, Madagascar, and the Red Sea.

Similar Indo-Pacific genera: Species of *Paraplexaura, Muricella, Anthogorgia, Trimuricea* and *Villogorgia* have a similar colony shape.

Remarks: Species of *Echinogorgia* appear in literature of the last century under such names as *Paraplexaura, Plexauroides, Thesea* and *Heterogorgia*.

(Echinogorgia, continued):

pyramid-like. Some of these sclerites are very large, like thorny blocks, and there may be spindles with thorn-like projections on one side. It is also not uncommon to find a few unusually large, tuberculate spindles here and there on the surface of the branches, seemingly distributed at random. Between and underneath these surface sclerites are small spindles, capstans and branched forms. The polyps are armed with generally large sclerites in a collaret and points arrangement. The point sclerites are usually hockey stick-shaped, and the collaret sclerites are bow-shaped. Sometimes there is only a single point sclerite below a tentacle, in which case it usually has a bifurcate base. Sclerites are usually red or colourless, occasionally yellow.

Colour: Red, brown, yellow, or white. Azooxanthellate.

Habitat and abundance: Can be common in fast-flowing turbid water, below depths of high irradiance.

Zoogeographic distribution: Because the characters of the genus have been confused, causing species to

Photos:
6, 7, 9, 13: Great Barrier Reef. *Photos: KF*
8, AND 10 - 12: Hong Kong. *Photos: KF*

7

8

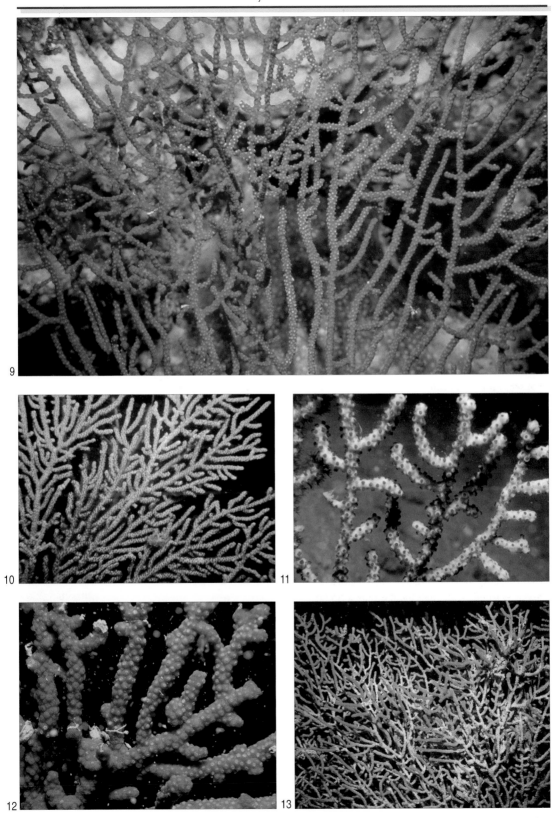

Menella

Gray, 1870

Colony shape: Commonly sparingly branched or un-branched and whip-like, but occasionally richly branched. Branches often quite thick. Side branches usually arising at near right angles from the main branches and curving upwards. Colonies do not form nets.

Polyps: Monomorphic, and totally retractile within low to prominent, hemispherical calyces.

Sclerites: The characteristic sclerite form is a leaf-scale which is shaped somewhat like a ping-pong bat that has the handle replaced by a tuberculate, bifurcate or complex "root" structure. The edge of the blade-like leaf may be relatively entire, or show various degrees of serration. In a few species, these leaf-scales are very small, and the blade portion of the scale is thickened, almost globe-like, with broad, irregular ribs down the sides. In all species the leaf scales cover the entire surface of the branches and form the walls of the calyces. Between and underneath the roots of the scales are small spindles, capstans, and branched forms. Crosses are common. The sclerites in the polyps are very variable in shape and number, ranging from none to very many. When present they do not form a proper collaret and points arrangement. Although a few sclerites may occur in the collaret position, most sclerites are longitudinally arranged. Portions of tripod-shaped sclerites may appear to circle the polyp neck below the points, but classical

bow-shaped collaret sclerites are rarely seen. Where polyp sclerites are few they are generally quite small, rod-like but rather flat. When numerous they tend to be quite large, not flat, and conspicuously ornamented with simple tubercles. Sclerites are red or colourless.

Colour: Purple, red, yellow, brown. Azooxanthellate.

Habitat and abundance: Fairly common, in particular in fast-flowing turbid water, below depths of high irradiance.

Zoogeographic distribution: Because the characters of the genus have been confused, causing species to appear in the literature under a number of different generic names, the range of the group is unclear, but it does appear to be wide spread. It is known to occur in New Caledonia, tropical and subtropical Australia, Papua New Guinea, some parts of Indonesia, Singapore, Andaman Islands, Malaysia, Sri Lanka, Bay of Bengal, Mauritius, Madagascar, South Africa, and the Red Sea.

Similar Indo-Pacific genera: Species of *Echinogorgia, Echinomuricea, Euplexaura, Bebryce* and *Paraplexaura* have similar colony shape.

Remarks: A number of species of *Menella* appear in the literature of the last century under the names *Echinogorgia* and *Plexauroides*.

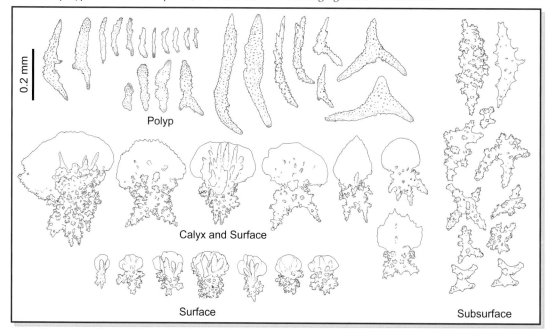

0.2 mm

Polyp

Calyx and Surface

Surface

Subsurface

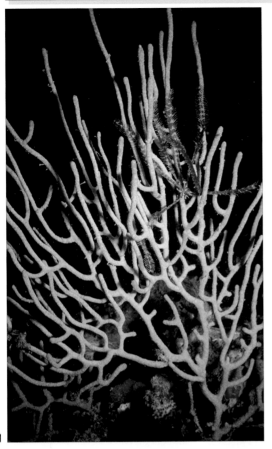

Photos:

1 - 3: Colonies from the Great Barrier Reef. *Photos: KF*

4: Hong Kong, ca 20 cm. *Photo: KF*

5: Preserved colony from the Great Barrier Reef. 25 cm. *Photo: KF*

Paraplexaura

Kükenthal, 1909

Colony shape: Commonly sparingly branched, but occasionally richly branched. Branches often quite thick. Side branches usually arising at near right angles from the main branches and curving upwards. Colonies do not form nets.

Polyps: Monomorphic, and totally retractile. Calyces generally low if they are present.

Sclerites: The surface of the branches contains sclerites with a complex tuberculate basal portion, and an upper surface covered in projections of various designs. In those species with the least complex forms, the upper surface of the sclerites is relatively flat, even slightly dished, and is covered in low, rounded mounds. In the most complex forms the upper surface of these sclerites has long thorns, sometimes branched. Numerous intermediate forms are encountered in different species. This type of sclerite can be relatively small, with only a couple of projections, but they are generally quite large, and sometimes massive. In some species, they can be quite narrow and cock's-comb-like, but all types and sizes may be found in any one colony. Between and underneath the bases of these sclerites are spindles, branched forms, capstans and capstan derivatives. Those species with the thorniest surface sclerites generally have the most developed calyces. The sclerites in the calyces are thick thornscales whose thorny projections reflect the design found in the surface sclerites. Those species in which the surface sclerites just have low mounds do not have calyces, but the rim of the polyp opening contains small, thick thornscales with rounded projections. Numerous intermediate states occur. The sclerites in the polyps are very variable in shape and number. There may be none at all, a few, or very many. When present they do not form a proper collaret and points arrangement. Although a few sclerites may occur in the collaret position, most sclerites are longitudinally arranged, and classical bow-shaped collaret sclerites are rarely seen. If polyp sclerites are few they are generally quite small, rod-like but rather flat. When numerous they tend to be quite large, not flat, and conspicuously ornamented with simple tubercles. Sclerites are red, orange or colourless.

Colour: Red, orange, or brown. Azooxanthellate.

Habitat and abundance: Uncommon, mostly encountered in fast-flowing turbid water, below depths of high irradiance.

Zoogeographic distribution: Because the characters of the genus have been confused, causing species to appear in the literature under a number of different generic names, the range of the group is unclear, but it does appear to be widespread. It is known to occur in New Caledonia, tropical and subtropical Australia, some parts of Indonesia, Singapore, Andaman Islands, Japan, and the Red Sea.

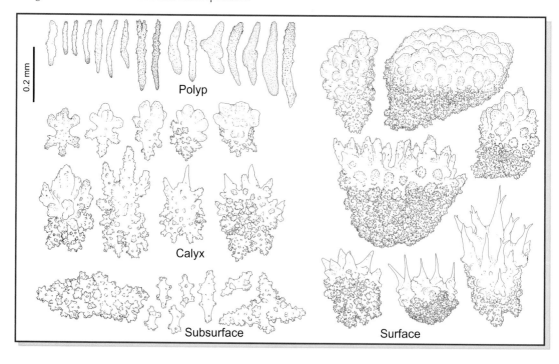

0.2 mm

Polyp

Calyx

Subsurface

Surface

1

2

3

4

Similar Indo-Pacific genera:
Species of *Menella,*
Echinogorgia, Echinomuricea,
and *Euplexaura* have similar
colony shapes.

Remarks: Species of
Paraplexaura appear in the
literature of the last century
under such names as *Echino-*
gorgia, Plexauroides and
Lapidogorgia.

Photos:

1, 3 AND 6: Colonies from the
Great Barrier Reef (GBR). Ca 20
cm. *Photos: KF*

2 AND 5: Hong Kong, view ca 12
cm **(2)** and 8 cm **(5)**. *Photos: KF*

4: Preserved specimen from the
GBR, ca 22 cm wide. *Photo: KF*

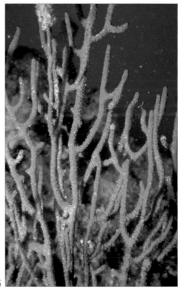

5

6

Astrogorgia

Verrill, 1868

Colony shape: Colonies grow in one plane as open fans, with irregular lateral branching. Branch joins are rare, and net-like fans are never formed.

Polyps: Monomorphic, retractile into low to tall calyces, but commonly preserved extended. Calyces are often in two rows, one down each edge of a branch, but can occur all around.

Sclerites: All surface sclerites in species of this genus are spindles. They can be thin, plump, short, long, curved or branched, and ornamented with anything from complex tubercles to small prickles. The tubercles are never arranged in more or less well defined girdles around the sclerite. On one colony the spindles can be all small, or a mixture of small to long. In some species the largest spindles are longer than the diameter of the branches. Smaller spindles, and short rods, often capstan-like, with 2 girdles of tubercles lie beneath the surface sclerites. The walls of the calyces are also formed from spindles, which are often arranged in eight double rows. In some species, however, they just lie in a longitudinal manner, and in others they lie transversally, in the direction of the branch. In some species the calicular sclerites are markedly smaller than those of the branch surface. The polyp head is always covered in numerous short spindles or rods arranged in eight longitudinal groups. Those at the base of the groups lie more obliquely and may become horizontal, many commonly continuing down onto the neck of the polyp, sometimes covering the whole neck region down to the rim of the calyx. The classical collaret formation of a few large bow-shaped sclerites does not occur. In the polyp tentacles there are usually short rods, of a different colour to the lower polyp sclerites and sometimes flattened, and there are often numerous granular scales. Sclerites can be red, orange, yellow or colourless.

Colour: Dark magenta, red, pink with white calyces, orange, yellow, cream, or yellowish brown. Azooxanthellate.

Habitat and abundance: Fairly common both in turbid and clear water.

Zoogeographic distribution: As a result of confusion regarding the correct characters of the genus, species have been described in the literature, often very poorly, under several different generic names, making the exact range difficult to determine. It is, however, wide spread, and certainly occurs in the Red Sea, Madagascar, Mauritius, Bay of Bengal, Sri Lanka, Andaman Islands, Malaysia, Singapore, Korea, Indonesia, tropical and subtropical Australia, Papua New Guinea, Japan, Funafuti, and New Caledonia.

Similar Indo-Pacific genera: Species of *Keroeides*, *Acanthogorgia*, *Anthogorgia*, *Muricella*, and *Bebryce*, can have similar growth forms.

(continued) ➡

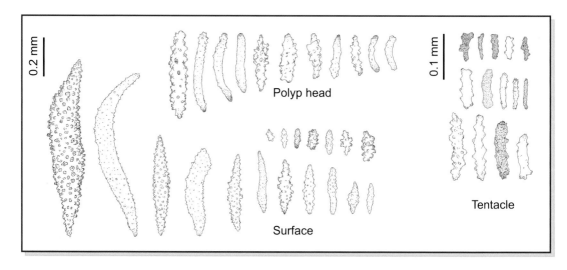

0.2 mm

0.1 mm

Polyp head

Tentacle

Surface

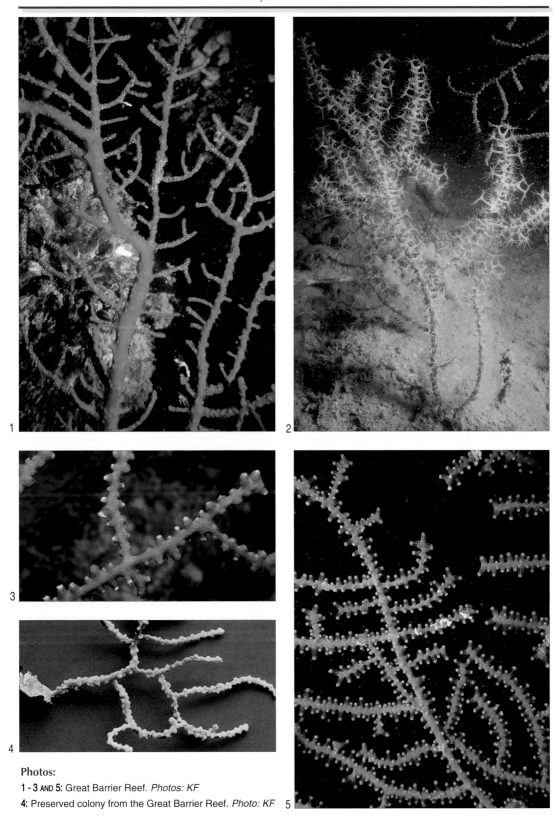

Photos:

1 - 3 AND **5:** Great Barrier Reef. *Photos: KF*

4: Preserved colony from the Great Barrier Reef. *Photo: KF*

(Astrogorgia, continued):

Remarks: The majority of the species described in the older literature under the generic name *Muricella* are actually species of *Astrogorgia*. The spindle-covered, non-retractile polyps of *Muricella* can easily be mistaken for the calyces of *Astrogorgia*. Species of *Astrogorgia* with spindles whose length exceeds the diameter of the branches have recently been grouped under the generic name *Acanthomuricea* to split the large group of species for convenience. But there is a large overlap between the two forms, and some species only have large spindles on some branches, so we prefer to include all species under the one genus.

Kükenthal established the genus *Anthoplexaura* in 1908 for a red gorgonian species that was thought to have dimorphic polyps. The author later realised that the supposed siphonozooids were actually the hydranths of a parasitic or symbiotic, burrowing, hydroid. Recently discovered colonies show the hydroid to be from the family Tubulariidae, and the gorgonian to be a species of *Astrogorgia*.

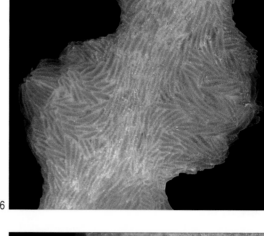

6

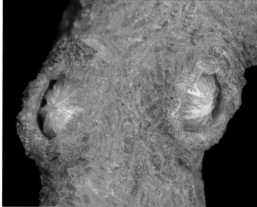

7

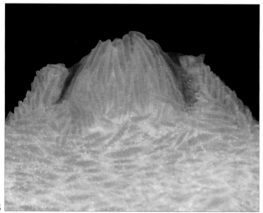

8

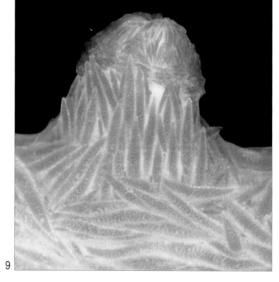

9

Photos:

6 AND 7: Calyces. *Photos: PA*

8: Polyp, with the front calyx wall removed. *Photo: PA*

9, 15 - 17: Various polyps, views ca 0.5 cm, showing the absence of the classical collaret-and points arrangement. *Photos: PA*

10 - 11: Northern Great Barrier Reef. *Photos: KF*

12: Southern Great Barrier Reef. *Photo: Gordon LaPraik*

13 - 14: Hong Kong, views ca 3 cm. *Photos: KF*

Rumphella

Bayer, 1955

Colony shape: Colonies, which can be up to a meter tall, grow as bushes. Branching may be sparse, resulting in a few, long whip-like branches, or very abundant, forming dense, shrub-like colonies. The branches are smooth and relatively thick, commonly with blunt, rounded tips. The holdfast is generally large and heavily calcified, and the coenenchyme is very thick.

Polyps: Monomorphic and retractile into the smooth branch surface.

Sclerites: The surface contains a layer of clubs. The club head has a terminal wart with a whorl of 3 large warts below. The subsurface layer contains spindles that have complex warts arranged in several girdles. The polyp head contains flattened rods arranged longitudinally in a group below the base of each tentacle. Sclerites are colourless.

Colour: Brown to greenish-grey with yellow to brown polyps. Zooxanthellate.

Habitat and abundance: Moderately common. In clear waters, it grows above 20 m depth, whereas in more turbid environments, it is restricted to shallow depths. Also found on wave-protected reef flats.

Zoogeographic distribution: Widespread, from southern Africa to the Solomon Islands.

Similar Indo-Pacific genera: *Hicksonella princeps, Isis hippuris.*

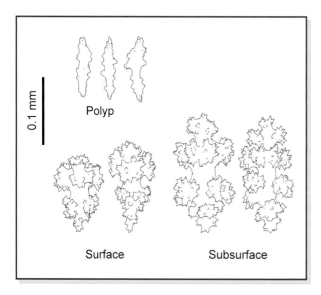

0.1 mm

Polyp

Surface   Subsurface

Photos:

ABOVE: Dried specimen from the northern Great Barrier Reef (GBR), ca 7 cm. *Photo: KF*

1, 2, 4, AND 5: GBR. *Photos: KF*

3 AND 6: Southern GBR, ca 20 cm. *Photos: Gordon LaPraik*

Hicksonella

Nutting, 1910

Colony shape: Only two species are known, and they have a completely different colony form, although both to varying degrees have the branchlets arising in an irregular pinnate manner from the major branches. In *Hicksonella princeps* the colonies generally have thick branches and form dense bushes. The coenenchyme may grow web-like right in the junction of some branches. In *Hicksonella expansa* the colonies are generally flabellate, as the coenenchyme forms extensive webbing. This results in colonies with leaf-like expansions that could be mistaken for algae or sponges. In both species, a small mound commonly surrounds the aperture of most of the retracted polyps, giving contracted colonies a slightly rough appearance. The coenenchyme is thick.

Polyps: Monomorphic, and retractile.

Sclerites: The surface contains a layer of clubs. The club head commonly has a terminal wart with a whorl of 3 large warts below, but irregular forms also occur. The subsurface layer contains spindles that may be covered with complex warts or have the warts arranged in several girdles. Also in the coenenchyme are found very long rods with a warty handle and a long, smooth or almost smooth, shaft. These long rods are unique to this genus, but their distribution in the coenenchyme does not seem to follow any detectable pattern. They may not appear in every sclerite sample taken for

analysis and can be easily overlooked, especially in *H. princeps*. Polyp heads contain flattened rodlets and scales arranged longitudinally in a group below the base of each tentacle. Sclerites are always colourless.

Colour: Greyish-brown. Zooxanthellate.

Habitat and abundance: Uncommon on the Great Barrier Reef, but can live in a wide range of habitats, generally above 15 m depth.

Zoogeographic distribution: Indonesia, Malaysia, the Philippines, Palau, northern Australia, the Great Barrier Reef, and New Caledonia.

Similar Indo-Pacific genera: *Hicksonella princeps* may be mistaken for *Rumphella*, and *Hicksonella expansa* looks very much the same as the leaf-like species of *Alertigorgia*.

Photos:
1: *H. princeps*, Great Barrier Reef (GBR), ca 20 cm. *Photo: KF*
2 AND 3: *H. expansa*. GBR, ca 15 - 20 cm. *Photos: Queensland Museum*
4: Dried specimen of *H. princeps*. GBR, ca 6 cm. *Photo: KF*
5: *H. expansa. Photo: Bioquatic Photo - A. J. Nilsen*
6: GBR. *Photo: Queensland Museum*
7: Sabah, Malaysia. *Photo: Frances Dipper*
8: Northern Territory, ca 20 cm. *Photo: PA*

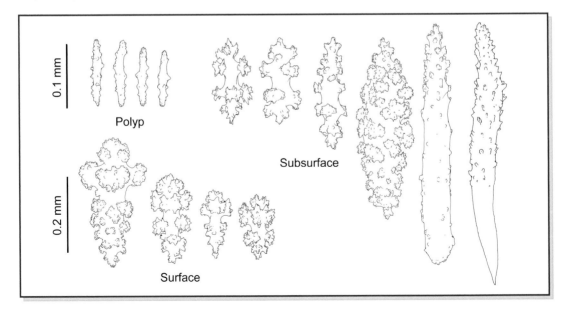

0.1 mm

Polyp

Subsurface

0.2 mm

Surface

Pinnigorgia

Grasshoff & Alderslade, 1997

Colony shape: Colonies are commonly richly branched, and bushy. The main stem, which can usually be traced through the colony, gives off numerous side branches from which many smaller branchlets arise more or less at right angles, and curving up. These branchlets may occur on two sides forming an irregular pinnate plume-like structure, or all around forming an irregular bottle-brush-like structure. In some habitats the small branchlets can be exceedingly long and thin, obscuring the pinnate or bottle-brush nature. Branch surfaces are usually smooth, though polyp apertures may be slightly domed.

Polyps: Monomorphic, retractile, and small.

Sclerites: In the polyp heads and tentacles there are small rod-like sclerites. In the surface of the stem and branches, sclerites of many shapes occur which are derived from simple spindles. They are commonly slightly curved, with complex warts on the concave side, and tall, well spaced, tooth-like or disc-like protrusions on the convex side. In some sclerites the protrusions are grouped at one end forming clubs. Sclerites are always colourless.

Colour: Beige to pale brown. Zooxanthellate.

Habitat and abundance: Relatively common on flow-exposed mid- and outer-shelf reef slopes below the reach of storm waves. Absent on turbid coastal reefs.

Zoogeographic distribution: Indonesia, Timor Sea, Great Barrier Reef, Micronesia and the Philippines.

Similar Indo-Pacific genera: Some colonies superficially resemble *Plumigorgia*. *Pinnigorgia* colonies with relatively neat and regular pinnate branching can also look similar to *Pseudopterogorgia*.

Remarks: In field guides this genus is commonly represented under the name of *Plexaura flava*, and also as *Lophogorgia*.

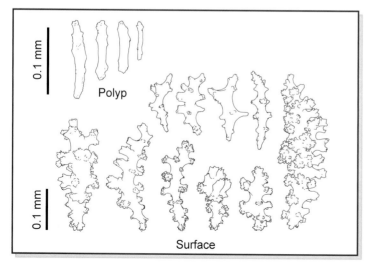

Photos:

Above: A dried, very bushy colony from the central GBR, ca 15 cm. *Photo: KF*

1: Central GBR, ca 20 cm. *Photo: KF*

2, 3 AND 4: Close-ups of colonies from the southern GBR. *Photos: KF*

5: An unusually droopy, large colony. Cartier Reef, north-west Australia. *Photo: Ann Storrie*

Pseudopterogorgia

Kükenthal, 1919

Colony shape: In their most recognisable form, colonies are composed of one or more neat, pinnately branched plumes. In some species the pinnate branching becomes very distant and irregularly spaced, and in others gross irregularities and richer branching tends to mask the pinnate tendency. A small mound may surround each polyp aperture when polyps are retracted.

Polyps: Monomorphic, small, and retractile, often within small mounds on the branch surface. The polyp aperture tends to be slit-like in the direction of the branch.

Sclerites: The surface contains straight and curved spindles with warts arranged in girdles. To a major or minor extent, the curved spindles have smaller warts on the convex side. Such curved sclerites with asymmetrically developed warts are called scaphoids. With Indo-Pacific species the differences in the size of the warts is minor. In the Caribbean species it is very marked, and the convex side is often smooth. In the polyp head, small, flattened rods with scalloped edges occur in chevrons (obliquely arranged double-rows) below the tentacle bases. Sclerites are mostly coloured.

Colour: Red, yellow, orange, brown, pink, violet, and white. More than 1 colour, usually only 2, may occur on a single colony, e.g. red and yellow, or pink and white. Azooxanthellate.

Habitat and abundance: Found in turbid environments and muddy estuaries. Rare on clear-water reefs of the GBR.

Zoogeographic distribution: Widespread from Mozambique to the central Indo-Pacific, but not yet reported in the western Pacific east of the Great Barrier Reef. *Pseudopterogorgia* is one of the most common gorgonian genera in the West Indies and other parts of the Caribbean.

Similar Indo-Pacific genera: Species of *Plumigorgia, Pinnigorgia, Pteronisis, Subergorgia,* and *Plumarella* form pinnate plumes, but other characters serve to easily distinguish them from *Pseudopterogorgia*.

Remarks: Scaphoids are often difficult to detect in a microscope preparation as they tend to lie with the convex side facing down, rather than on their sides, and so appear as straight spindles. Therefore the sclerites have to be rolled to observe any curvature. This genus is very similar to *Lophogorgia*, which occurs in the Atlantic, and further research is needed. Also, most of the Caribbean species of *Pseudopterogorgia* seem to be zooxanthellate and may represent a different genus.

0.1 mm

Polyp

Surface

Photos:

1: Northern Territory, Australia, ca 25 cm. *Photo: PA*

2: Southern GBR, ca 10 cm. *Photo: Gordon LaPraik*

3: Dried specimen from a coastal reef of the central GBR, ca 15 cm. *Photo: KF*

Guaiagorgia

Grasshoff & Alderslade, 1997

Colony shape: Untidy, straggling, very sparsely branched colonies formed from long, thin branches commonly arising more or less at right angles from the stem and other branches. Unbranched colonies have also been recorded. All branches are of a similar diameter, and polyp apertures appear as dark dots in preserved specimens.

Polyps: Monomorphic, with long bodies when expanded, and completely retractile.

Sclerites: The surface contains narrow and plump spindles, occasionally a few clubs, with the warts arranged irregularly or in girdles. The polyp head has small rods in a loose collaret and points arrangement. Colourless.

Colour: The only live colour recorded so far is blue. Possession of zooxanthellae is unknown.

Habitat and abundance: Rare. On flat bottom in sand and on rock, or on slight inclines.

Zoogeographic distribution: Indonesia, northern Australia and New Caledonia.

Similar Indo-Pacific genera: Some species of *Menella* and *Echinomuricea* have similar growth forms, but conspicuously different sclerites serve to easily distinguish them from *Guaiagorgia*.

Photos:

1 AND 2: Colony of unknown origin in a reef tank. *Photos: Julian Sprung*

3: Dried colony from New Caledonia. Scale bar represents 5 cm. *Photo: Manfred Grasshoff*

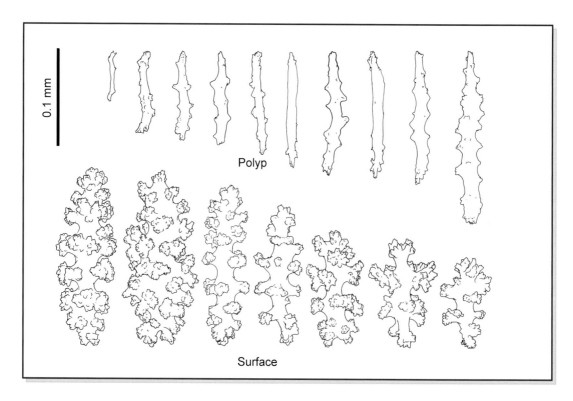

0.1 mm

Polyp

Surface

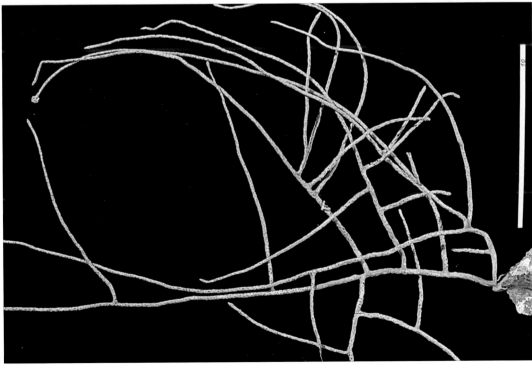

Ellisella

Gray, 1858

Colony shape: Richly or sparsely branched, generally bushy colonies, up to at least a meter high, formed by repeated dichotomous branching. Branches are often long and whip-like.

Polyps: Monomorphic, and very contractile but not retractile. They may be arranged all around the branches or occur in a few longitudinal tracts. The polyps commonly contract and bend over to form mounds on the surface of the branches, which may be small or very pronounced and lobe-like.

Sclerites: Double heads and waisted spindles. The spindles are slightly longer, to much longer, than the double heads. Surface sclerites are usually coloured, while the subsurface sclerites from coloured colonies can be colourless.

Colour: Red to orange, pink, white, orange-yellow with red polyps, or red with white polyps. Azooxanthellate.

Habitat and abundance: Occurs in both tropical and temperate waters, clear and turbid, but is rare above 5 m depth. Common on near-shore reefs of the Great Barrier Reef on current-swept muddy bases of reefs.

Zoogeographic distribution: Widespread, and appears to occur in most areas covered by this book. Also occurs in the Atlantic and the Mediterranean.

Similar Indo-Pacific genera: Species of *Dichotella* have a similar growth form, but are easily distinguishable by their sclerites. Isolated branch fragments of *Ctenocella*, *Viminella* and *Ellisella* cannot be separated by an examination of the sclerites.

Photos:

BELOW, AND 2: Smooth, 30 cm tall colonies from the northern GBR. *Photos: KF*

1: Papua New Guinea. *Photo: Roger Steene*

3: Southern GBR. *Photo: KF*

4: Solomon Islands. *Photo: Roger Steene*

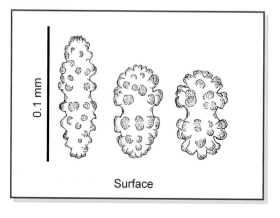

0.1 mm

Surface

Viminella

Gray, 1870

Colony shape: Colonies up to 2 meters tall and whip-like, uncommonly with one or a few side branches.

Polyps: Monomorphic, and very contractile but not retractile. They may be arranged all around or occur in a few longitudinal tracts. The polyps commonly contract and bend over to form mounds on the surface of the branches, which may be small or very pronounced and lobe-like.

Sclerites: Double heads and waisted spindles. The spindles are slightly longer to much longer than the double heads. Surface sclerites are usually coloured, while the subsurface sclerites from coloured colonies can be colourless.

Colour: Red to orange, pink, white, yellow, orange-yellow with red polyps, or red with white polyps. Azooxanthellate.

Habitat and abundance: Uncommon. Found in coastal areas in current-swept muddy bases of reefs. Sometimes found lying free and unattached on sandy bottoms.

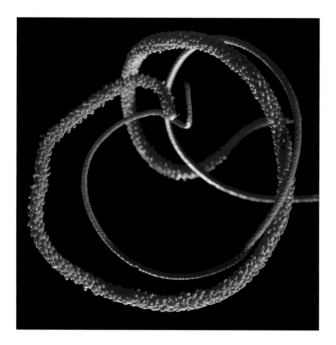

Zoogeographic distribution: Widespread, so could reasonably be expected to be found in most areas covered by this book. Also occurs in the Atlantic and the Mediterranean.

Similar Indo-Pacific genera: Species of *Junceella* have the same unbranched colony form but are easily distinguishable by an examination of the sclerites. Solitary branch fragments of *Ctenocella, Viminella* and *Ellisella* cannot be separated by an examination of the sclerites.

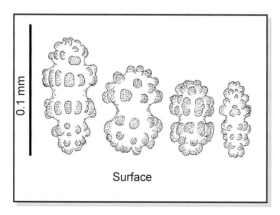

0.1 mm

Surface

Photos:

ABOVE: Preserved colony from New Caledonia. *Photo: Manfred Grasshoff*

1: Papua New Guinea. *Photo: Gary Williams*

2: Guam. *Photo: Gustav Paulay*

3: Southern GBR. *Photo: Gordon LaPraik*

4: Northern GBR. *Photo: KF*

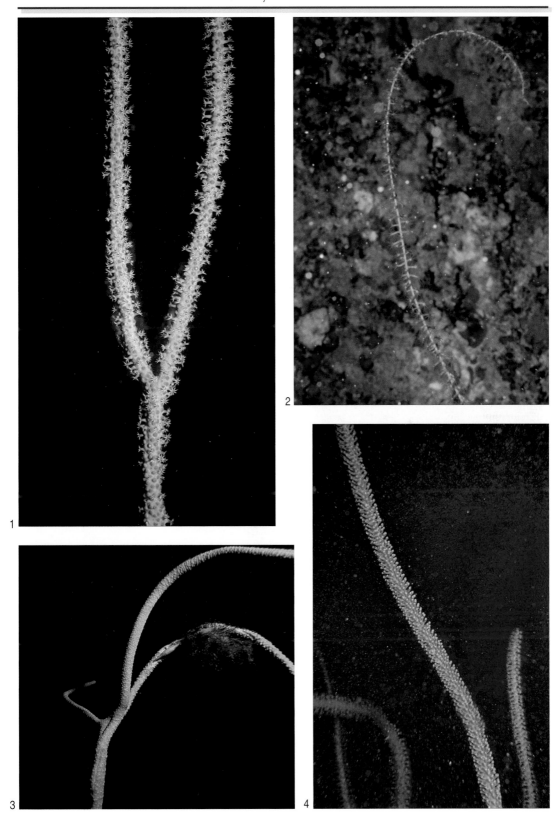

Ctenocella

Valenciennes, 1855

Polyps: Monomorphic, and very contractile but not retractile. The polyps commonly contract and bend over to form small mounds on the surface of the branches.

Sclerites: Double heads and waisted spindles. The spindles are never much longer than the double heads. Usually coloured.

Colour: Red to orange, brown to yellow, and white. Azooxanthellate.

Habitat and abundance: In northern Australia and the Great Barrier Reef, it is common in current-swept turbid waters of near-shore environments, on the base of reefs and in inter-reefal areas. Rare above 5 m depth.

Zoogeographic distribution: *Ctenocella pectinata* is considered to be the only valid species. It has been reported from Indonesia, Singapore, Thailand, northern Australia and the Great Barrier Reef.

Similar Indo-Pacific genera: Confusion could exist between young colonies of *Ctenocella* and any aberrantly branched *Ellisella* such as *Ellisella eustala*. Branch fragments of *Ctenocella*, *Viminella* and *Ellisella* cannot be separated by an examination of the sclerites.

Colony shape: In the most characteristic form, colonies are comb- or lyre-shaped, with thin parallel branchlets arising vertically from two opposing, upward curving main branches. In modified forms there may be just a single comb on one side, or several overlapping combs. Colonies can grow more than 1.5 meters across.

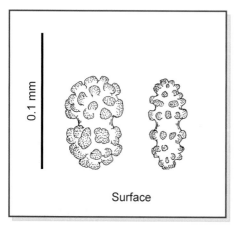

0.1 mm

Surface

Photos:

ABOVE: Great Barrier Reef (GBR). *Photo: Roger Steene*

1 AND **4:** GBR. *Photos: KF*

2: Central GBR. *Photo: Götz Reinicke*

3: A predatory snail (*Aclyvola lanceolata*) on a *Ctenocella* branch. Southern GBR, ca 5 cm. *Photo: Gordon LaPraik*

5: On turbid, coastal reefs of the GBR, fans can grow over 1.5 m wide. Central GBR. *Photo: KF*

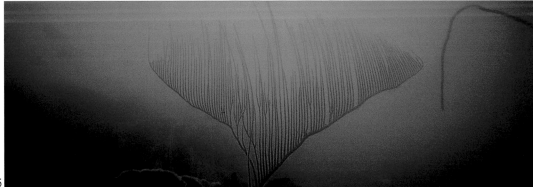

Junceella

Valenciennes, 1855

Colony shape: Unbranched, whip-like colonies that can grow over 2 meters tall.

Polyps: Monomorphic, and very contractile but not retractile. They may be arranged all around or in a few longitudinal tracts, and when contracted they are generally angled upwards. Contracted polyps may be very pronounced and look like curved snake heads, they may just form small mounds, or they may appear as small scales pressed against the surface of the branches or into a small depression in the branch

Sclerites: The surface contains clubs. In by far the majority of species the club heads are well formed from a cluster of distally pointed tubercles. In other species the tubercles are poorly formed and the club head looks water-worn, and globular with a lumpy surface. The sublayer contains symmetrical, or near symmetrical capstans. Surface sclerites are usually coloured, while the subsurface sclerites from coloured colonies can be colourless.

Colour: Many variations from red, through orange, yellowish brown, beige, grey, yellow, to white. Azooxanthellate.

Habitat and abundance: Common in coastal waters of the Great Barrier Reef in current-swept muddy bases of reefs, and moderately common on mid-shelf reefs at greater depth. Uncommon on outer-shelf reefs. Also occurs in turbid coastal inter-reefal environments and muddy estuaries.

Zoogeographic distribution: Red Sea, South China Sea, central Indo-Pacific, Great Barrier Reef, Micronesia, eastwards at least as far as New Caledonia.

Similar Indo-Pacific genera: Species of *Viminella* have the same unbranched colony form, but can be easily told apart by examination of the sclerites. An unbranched fragment of *Dichotella* would be very difficult to distinguish from *Junceella* using sclerite analysis

Remarks: In some species at least, the colony tip can detach from the parent as a means of asexual reproduction, fall to the sea floor, and re-attach.

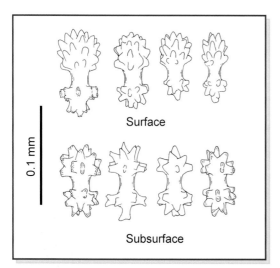

Surface

0.1 mm

Subsurface

Photos:

1: Sinai, Red Sea. *Photo: Günther Bludszuweit*

2: *Junceella* with commensal goby (*Cottogobius yongei*). Southern Great Barrier Reef (GBR). *Photo: Gordon LaPraik*

3: *Junceella* with predatory snail (*Phenacovolva rosea*). Southern GBR. *Photo: Gordon LaPraik*

4 AND 5: Southern GBR. *Photos: Gordon LaPraik*

6: Central GBR. *Photo: KF*

7: Northern GBR. *Photo: KF*

Dichotella

Gray, 1870

Colony shape: Abundantly or sparsely branched, bushy or planar colonies, up to at least 1 meter high, formed by repeated dichotomous branching. Branches are usually relatively short.

Polyps: Monomorphic, and very contractile but not retractile. They may be arranged all around or in a few longitudinal tracts, and when contracted are generally angled upwards. Contracted polyps may be very pronounced and look like curved snake heads, they may just form small mounds, or they may appear as small scales pressed against the surface of the branches or into a small depression in the branch.

Sclerites: The surface of the coenenchyme contains clubs where the head is formed from a cluster of distally pointed tubercles. The subsurface contains symmetrical, or near symmetrical capstans. Surface sclerites are usually coloured, while the subsurface sclerites from coloured colonies can be colourless.

Colour: Many variations from red, through orange, yellowish brown, yellow, to white. Azooxanthellate.

Habitat and abundance: Common in coastal environments from very muddy estuaries to clearer areas on current-swept bases of reefs. Occasionally found on clear-water reefs, but rarely encountered above 5 m depth.

Zoogeographic distribution: *Dichotella gemmacea* is considered to be the only valid species. It has been reported from Central Indo-Pacific, South China Sea, eastwards at least as far as New Caledonia.

Similar Indo-Pacific genera: Species of *Ellisella* have the same colony form but are easily distinguishable by an examination of the sclerites. An unbranched fragment of *Dichotella* would be very difficult to distinguish from *Junceella* using sclerite analysis.

Photos:
BELOW: Dried colony, GBR, ca 30 cm wide. *Photo: KF*
1: Southern GBR, ca 40 cm. *Photo: Gordon LaPraik*
2: Solomon Islands, ca 60 cm. *Photo: Gary Williams*
3 AND 5: Close-ups of colonies from the GBR. *Photos: KF*
4: Torres Straits, northern Australia. *Photo: KF*
6: Preserved colony from PHOTO 4. View ca 13 cm. *Photo: KF*

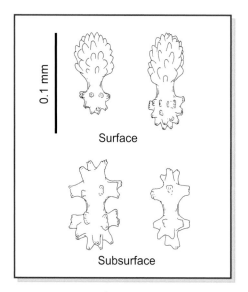

0.1 mm

Surface

Subsurface

Heliania

Gray, 1860

Colour: Red to orange, orange-yellow, white. Azooxanthellate.

Habitat and abundance: Would seem to be quite rare at diving depths. The shallowest record is 25 m, but usually the genus is found much deeper, even down to 600 m.

Zoogeographic distribution: The Maldives, Christmas Is., Moluccas, Philippines, New Caledonia and Japan.

Similar Indo-Pacific genera: Some colonies may be similar to some species of *Nicella*.

Colony shape: Flabellate, planar colonies, often richly branched. Notably, small branches generally arise from just one side of the thicker branches. Branch joins are sometimes present.

Polyps: Monomorphic, and very contractile but not retractile. They can be arranged all around or occur in one or more longitudinal tracts. The contracted polyps form small mounds on the branches.

Sclerites: The surface layer of the coenenchyme contains clubs in which the head is formed from a cluster of distally pointed tubercles. The subsurface contains symmetrical, or near symmetrical, spindle- or rod-like forms with a narrow waist. Capstan-like forms are very uncommon. Sclerites are coloured.

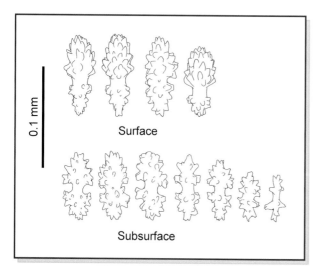

Photos:

ABOVE, AND 2: Preserved colony from New Caledonia; the scale bar in the large picture represents 5 cm. *Photo: Manfred Grasshoff*

1: Maledives. *Photo: Harry Erhardt*

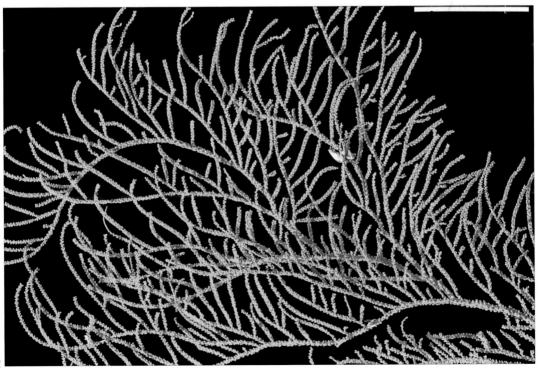

Verrucella

Milne Edwards & Haime, 1857

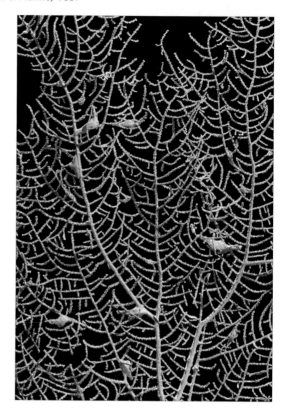

Colony shape: Colonies are up to at least 50 cm tall, richly branched in one plane or bushy and compressed, with short branches. The number of branch joins is extremely variable, ranging from colonies in which every terminal branch is free, to those which form complete net-like fans.

Polyps: Monomorphic, and very contractile but not retractile. The contracted polyps generally form small mounds on the surface of the branches.

Sclerites: Double heads, and both narrow and plump spindles. The narrower and the shorter spindles usually have a waist. Surface sclerites are coloured, while the subsurface sclerites can be almost colourless.

Colour: Many variations from red, through orange, yellowish brown, to yellow. Azooxanthellate.

Habitat and abundance: Uncommon. Has been reported from turbid and clear-water environments.

Zoogeographic distribution: Widespread, and is likely to be found in most areas covered by this book.

Similar Indo-Pacific genera: Some species with sparse branch fusion may resemble *Nicella* or *Jasminisis*.

Remarks: The genus *Umbracella* has recently been synonymised under *Verrucella*.

0.1 mm

Surface

Photos:

ABOVE, AND 6: Dried specimen from New Caledonia, ca 30 cm, and detail thereof. *Photos: Manfred Grasshoff*

1: Hong Kong, ca 12 cm. *Photo: KF*

2: Detail of **1**, ca 3 cm. *Photo: KF*

3: Dried specimen from New Caledonia, ca 20 cm. *Photo: Manfred Grasshoff*

4: Detail of **3**, ca 5 cm. *Photo: Manfred Grasshoff*

5: Northern GBR, ca 15 cm. *Photo: KF*

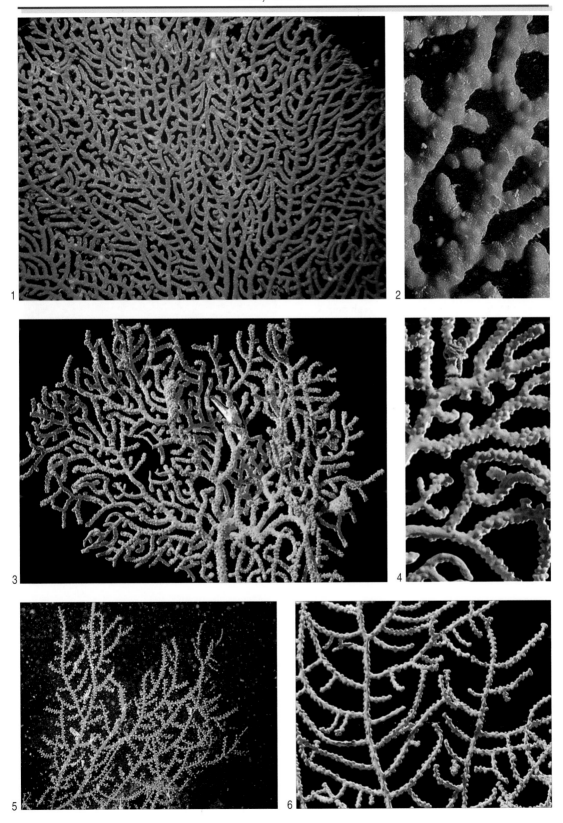

Nicella

Gray, 1870

Colony shape: Colonies are up to about 50 cm tall, and sparingly to moderately branched in one plane. Specimens commonly show a tendency towards dichotomous branching, made irregular by out-of-place lateral branches. Colonies do not form nets. Polyps may be very conspicuous.

Polyps: Monomorphic, very contractile but not retractile, and arranged all around the branches or biserially. Contracted polyps may appear as small or large mounds, but in some species they are very tall and volcano-shaped.

Sclerites: The surface layer contains small double heads, above a thick subsurface layer of flattened rods or spindles generally without a distinct waist. These subsurface layer sclerites can be two to four times the length of the double heads, but are the same size in a few species. Amber, yellow or colourless.

Colour: White or brown. Azooxanthellate.

Habitat and abundance: Rare. Has been found at diving depth in New Caledonia and in Palau, but appears to be generally rare at depths less than 50 m.

Zoogeographic distribution: All tropical seas.

Similar Indo-Pacific genera: Some species resemble *Verrucella*, others may be similar to *Heliania*. Some dry colonies resemble *Anthogorgia*.

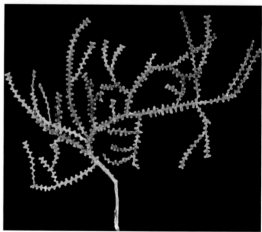

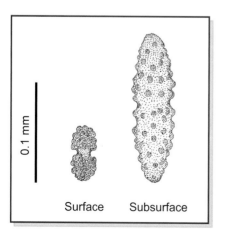

Surface   Subsurface

Photos:

Top: Preserved specimen, 35 cm tall. *Photo: PA*

Above: Preserved specimen, 17 cm across. *Photo: PA*

1: Palau. *Photo: Coral Reef Research Foundation*

2: Two preserved species from New Caledonia; the scale bar represents 5 cm. *Photo: Manfred Grasshoff*

1

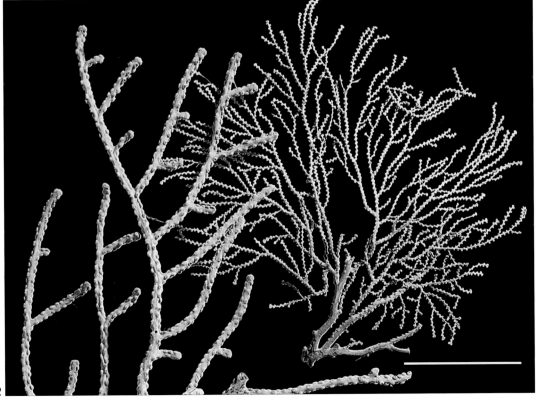

2

Ifalukella

Bayer, 1955

Colony shape: Colonies are fan-like or bushy and up to 15 cm tall, arising from a massive calcareous holdfast. The axis is spirally ridged, and the ridges extend onto the holdfast as high crests. The branching pattern is irregularly lateral and the branches are very fine, short, and brittle. More than one colony may originate from the same holdfast. The coenenchyme is very thin.

Polyps: Monomorphic, very small, and retractile into low domes.

Sclerites: Very small, flattened, peanut-shaped structures with a coarse crystalline surface.

Colour: Greenish to beige to brown. Zooxanthellate.

Habitat and abundance: Unknown.

Zoogeographic distribution: Known only from Ifaluk Atoll in the central Caroline Islands, from Chuuk in Micronesia, and from Guam in the Mariana Islands.

Similar Indo-Pacific genera: None.

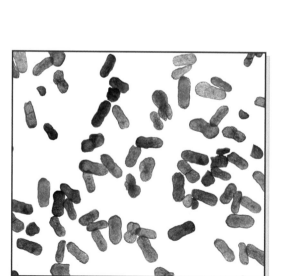

Photo:
TOP, AND ABOVE: Preserved specimen from Guam. *Photos: PA*
LEFT: Sclerites of *Ifalukella* from a Guam specimen (0.025 - 0.09 mm in length). *Photo: PA*
1, 3, AND 4: Guam. *Photos: Gustav Paulay*
2: Chuuk, Micronesia. *Photo: Coral Reef Research Foundation*

Plumigorgia

Nutting, 1910

Colony shape: Colonies are pinnate, often richly branched, and planar to compressed or bushy. They grow up to at least 20 cm tall. Branchlets may be very fine and somewhat flattened, to thick and cylindrical. Branchlets are usually short, but can occur long and thread-like in some specimens of one species. The coenenchyme is often expanded web-like between the bases of adjacent branchlets. Species with few sclerites may feel rubbery and slimy. In preserved specimens and on some branches of live colonies the axis may protrude at the tips of the branchlets. Colonies have a small calcareous holdfast. The coenenchyme may be very thin or very thick.

Polyps: Monomorphic, very small, distributed all around or biserially, and retractile, leaving the colony surface flat or raised into very small domes.

Sclerites: Very small structures with a coarse crystalline surface, occurring in the shape of ovals, sub-spheroids, peanuts, crosses, multiradiate stars, and irregular shapes. In one species sclerites are only found on the tips of the branchlets, and can be absent altogether. Colourless.

Colour: Pale greyish-pink to grey brown with violet hues. Probably zooxanthellate.

Habitat and abundance: Common on mid- and outer-shelf reefs of the Great Barrier Reef, in shallow, flow-exposed environments.

Zoogeographic distribution: Common on the Great Barrier Reef, reported several times from Micronesia and Indonesia, and once from Ibugos Island, China Sea.

Similar Indo-Pacific genera: Some light grey colonies superficially resemble *Pinnigorgia*.

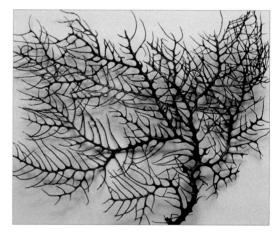

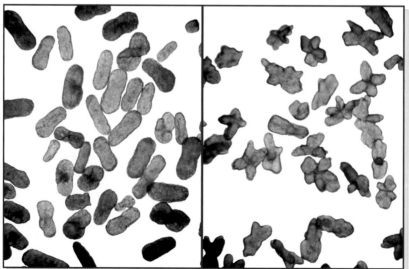

Photos:

ABOVE: A dried specimen of *P. schubotti*, display-ing the distinct webbing. Great Barrier Reef, ca 12 cm. *Photo: KF*

LEFT: Two types of sclerites found in *Plumigorgia*. *Photos: PA*

1 -3, AND 5: Colonies from various locations on the GBR. *Photos: KF*

4: *Photo: Roger Steene*

Plumarella

Gray, 1870

Colony shape: Pinnately branched colonies with one or many plumes that grow in one plane. The branchlets may be very slender and close together forming feather-like structures. The axis commonly has a metallic sheen.

Polyps: Monomorphic, very small, and contractile but not retractile. They are arranged biserially, and are curved to face obliquely along the branch.

Sclerites: The polyp body is covered in 8 rows of scales that may have an even or irregular outline. When the polyp contracts and the tentacles fold in over the mouth, the summit becomes covered by an operculum of 8 triangular or triradiate scales. The uppermost, or marginal, body scales surrounding the operculum may have a projecting point. The surface of the branches between the polyps has a layer of plates or flattened spindles. Always colourless.

Colour: The axis is often orange-golden. Because sclerites and coenenchyme are colourless, the colony colouration depends on the extent that the metallic sheen of the axis is visible through the surface tissue. Azooxanthellate.

Habitat and abundance: In northern Australia, *Plumarella penna* is not uncommon and grows in turbid environments with strong current. Off Australia's North West Shelf it occurs in clearer water below 60 m.

Zoogeographic distribution: Known from deep water in many of the world's temperate and tropical oceans. Only one species, *Plumarella penna*, has been recorded from shallow water (less than 10 m), and only from Darwin, northern Australia. This species has been collected, using dredges, off north-western Australia, and also near some of the islands of eastern Indonesia where in all probability it will be found at diveable depths.

Similar Indo-Pacific genera: *Plumarella penna* has extremely small polyps and a particularly fine branch structure that could cause it to be confused with some of the larger plumate hydroid species.

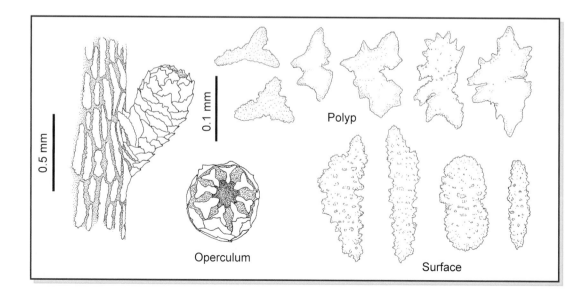

Operculum   Polyp   Surface

Photos:

1: Preserved colony, 16 cm wide. *Photo: PA*

2: A colony from Darwin, northern Australia, *Photo: Karen Gowlett-Holmes*

LEFT AND 3: Details of the colony shown in 1. *Photos: PA*

Stephanogorgia

Bayer & Muzik, 1976

Colony shape: Mostly small, delicate, profusely branched colonies that grow in one plane. The main branches grow in a zigzag manner, alternately giving off side branches at each bend. These thinner secondary branches may remain unbranched, but generally they re-branch in a similar zigzag manner. The basal holdfast is calcified, and the metallic golden coloured branch axis can be seen beneath the extremely thin coenenchyme.

Polyps: The polyps are very small, angled upwards, and arranged in a single row or 2 opposing rows along each branch.

Sclerites: The dominant sclerites are very small, extremely thin, flat scales, and are difficult to detect. They may be long and narrow or short and broad, and commonly have a waist. They occur primarily on the polyp bodies, and may be sparse or abundant, spread over the whole body, or arranged in 8 groups converging at each tentacle base. When abundant they may encroach onto the branch surface in the vicinity of the polyp base. In the tentacles, there are sometimes minute narrow rods with small cones or serrated ridges on the surface. All sclerites are colourless.

Colour: White or yellow colonies are recorded. Azooxanthellate.

Habitat and abundance: Locally common in deeper clear water under overhangs and ledges.

Zoogeographic distribution: Known only from Bali, Solomon Islands, Palau, Chuuk, Fiji, and Jolo Island in the Sulu Archipelago.

Similar Indo-Pacific genera: White colonies of *Zignisis* would look identical, but *Stephanogorgia* does not have a segmented axis.

Photos:
1: Solomon Islands. *Photo: Gary Williams*
2: Bali, Indonesia. *Photo: Fred Bavendam*
3: Palau, Micronesia. *Photo: Coral Reef Research Foundation*

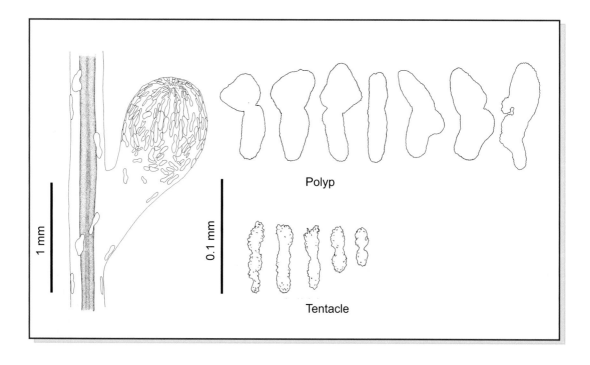

1 mm

0.1 mm

Polyp

Tentacle

Isis

Linnaeus, 1758

Colony shape: Branches are always smooth and cylindrical. A very thick coenenchyme covers the characteristic, alternating, white and brown-black segmented axis, which does not contain sclerites. The white calcareous internodes are conspicuously ridged. The branching pattern of colonies is extremely variable. Colonies can be sparingly or richly branched, and although most grow fan-like in a single plane, they can be extremely bushy. Most branches are usually quite short, but colonies with long whip-like branches are also known. Branches tend to arise approximately at right angles to the major branches, and although they may continue to grow in this direction it is more common for them to curve upwards, sometimes forming numerous candelabrum-like elements within a colony.

Polyps: Monomorphic, small, densely arranged all around the branches, and completely retractile within the coenenchyme leaving the branch surface smooth.

Sclerites: The surface layer contains small clubs, with a whorl of 3 large warts surrounding a terminal wart on the head. Below there is a subsurface layer of warty sclerites of numerous forms: 6- to 8-radiate capstans, spindles, ovals, and spheroids. The variation in size and shape between colonies from different environments and geographic regions is very large. Always colourless.

Colour: Bright yellow to green or brown. The colour of the polyps is similar to that of the coenenchyme. Zooxanthellate.

Habitat and abundance: Common and widely distributed in shallow moderately clear waters. On the Great Barrier Reef, it is particularly abundant on mid-shelf reefs, where it occurs in shallow water away from wave action. Rare in turbid coastal areas.

Zoogeographic distribution: *Isis hippuris* is the only recognised species. It has been recorded from the Andamans, Philippines, Taiwan, Palau, Indonesia, Papua New Guinea, Great Barrier Reef, and the Ryukyu Islands.

Similar Indo-Pacific genera: *Rumphella* forms similarly bushy, stout colonies in shallow water. The tendency for branches to arise at right angles and curve upwards generates the same colony form that is found in species of *Euplexaura*.

Photos:

BELOW: Close-up of the segmented axis. Length: ca 2 cm. *Photo: PA*

1 - 5: Great Barrier Reef. *Photos: KF*

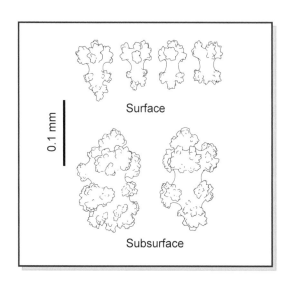

Surface

0.1 mm

Subsurface

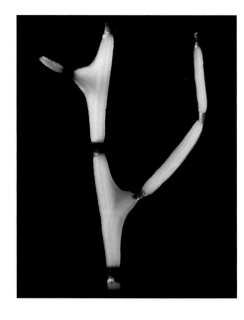

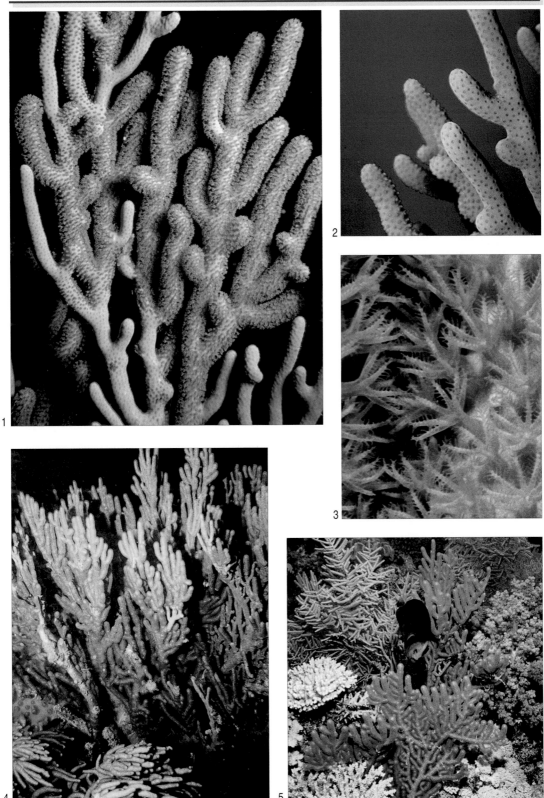

Jasminisis

Alderslade, 1998

Colony shape: Colonies are usually very small, but two temperate water species grow to about 20 cm. Branching is more or less dichotomous, resulting in planar colonies that are sometimes lyrate. The axis is segmented, does not contain sclerites, and branches usually commence with an axial node. The axial internodes on the thinner branches are generally four-sided and covered with blunt spines. In the thicker branches and the stem, they have multiple ridges and small spines.

Polyps: Monomorphic, not retractile, arranged all around or biserial, and generally curved upwards so the polyp mouth faces along the branch. In one temperate water species, the contracted polyps are seated in depressions in the branch surface and give the false impression that they are retracted.

Sclerites: Both the head and body of the polyp is covered in scales, mostly crescent-shaped, with leafy or tooth-like projections. In some species the basal edge of the scales has long root-like structures. The scales of the polyp head and body are separated by a naked neck zone, but in preserved specimens this is usually closed and difficult to see. When the polyp contracts and the tentacles fold in over the mouth, the summit becomes covered by an 8-segmented operculum-like structure. Each segment has a terminal triangular or triradiate scale and 1 - 2 preceding crescentic scales. In the only warm water species described so far (*J. cavatica*), only the 8 tri-angular scales fold over the mouth to any extent. The polyp tentacles contain small boomerang- or butterfly-like scales. The sclerites of the coenenchyme can include capstans, spheroids, and spindles (occasionally branched), all unilaterally developed with leafy or tooth-like projections. Sclerites are always colourless.

Colour: *Jasminisis cavatica*, which is yellow, is the only tropical species in which the live colour is known. In all species the calcareous axial internodes are white, and the horny nodes are brown to yellow. Azooxanthellate.

Habitat and abundance: *Jasminisis cavatica* is rare, only being known from two sites, and was found growing down from the roof of shallow caves at about 10 m depth. The other three described species are from temperate waters at 64 to 102 m depth, and an as yet undescribed species has been found at 75 m depth in Papua New Guinea.

Zoogeographic distribution: *Jasminisis cavatica* is only known from the southern and central Great Barrier Reef. The other species were dredged off the south-eastern coast of Australia, and a new species has been found in Papua New Guinea.

Similar Indo-Pacific genera: The growth form of *Jasminisis cavatica* is the same as some species of *Verrucella*.

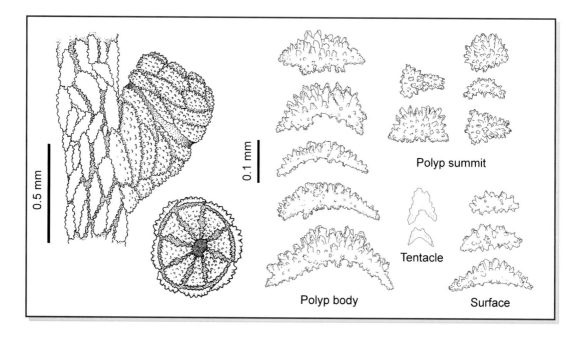

0.5 mm

0.1 mm

Polyp summit

Tentacle

Polyp body

Surface

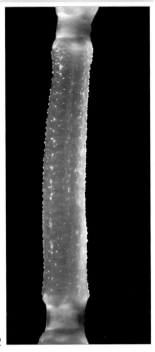

Photos:

1: An as yet undescribed species from Papua New Guinea, photographed at 75 m depth. *Photo: Coral Reef Research Foundation*

2: Close-up of an axial internode. *Photo: PA*

3: Preserved colonies, the left one being 7.3 cm across. *Photo: PA*

Pteronisis

Alderslade, 1998

Colony shape: Colonies are small, and usually profusely branched in a pinnate (feather-like) manner, growing in one plane. The calcareous internodes are generally 4-sided near the tips of the narrow twigs, but have more than 4 sides nearer to the twigs' base. These internodes also have small denticles, which can occur over the whole internode but are usually just found on the end shoulders. In the stem and the thicker main branches, the internodes have longitudinal ridges, and generally no denticles. *Pteronisis provocatoris*, the only warm water form, is highly variable in shape, and can be found as closely pinnate, vaguely pinnate, or sparsely branched colonies, and all forms in between.

Polyps: Monomorphic, not retractile, arranged all around or biserial, and curved upwards so the mouth faces along the branch.

Sclerites: Both the head and body of the polyp are covered in scales, mostly crescent-shaped, with leafy or tooth-like projections. In some species the basal edge of the scales has long root-like structures. The scales of the head and the body are continuous and not separated by a naked neck zone. When the polyp contracts and the tentacles fold in over the mouth, the summit becomes covered by an 8-segmented operculum-like structure. Each segment has a terminal triangular or triradiate scale and 1 - 2 preceding crescentic scales. The polyp tentacles contain small boomerang- or butterfly-like scales. The sclerites of the coenenchyme usually include ovals and spindles (occasionally branched), all unilaterally developed with leafy or tooth-like projections. Sclerites are always colourless.

Colour: *P. provocatoris* is rose, rose-orange, or brownish orange when alive. Other temperate water species are described as being greyish orange, greyish red, and brilliant red. All species are white or yellowish-white when preserved. Azooxanthellate.

Habitat and abundance: *Pteronisis provocatoris* is common and grows in both sheltered and wave-exposed coral reef zones in about 40 - 50 m of water.

Zoogeographic distribution: *Pteronisis provocatoris* is known only from New Caledonia. All other species are known only from deep water off the south, south-western and south-eastern coasts of Australia.

Similar Indo-Pacific genera: Closely pinnate species of *Pseudopterogorgia* could have a similar colony form, but they have different sclerites, and do not have a segmented axis.

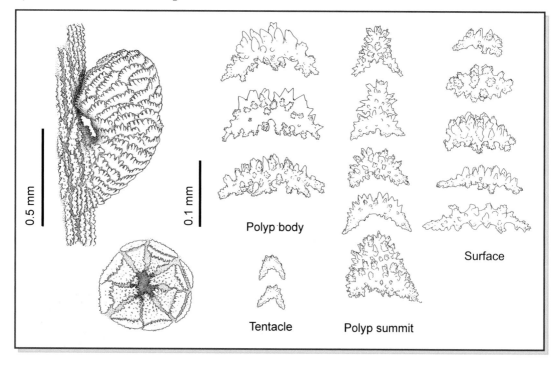

0.5 mm

0.1 mm

Polyp body

Surface

Tentacle

Polyp summit

1

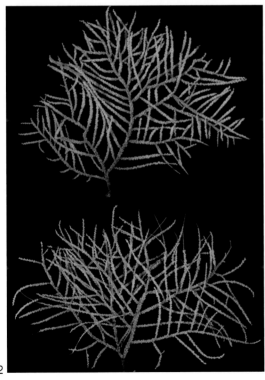

2

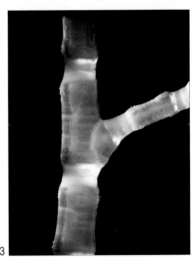

3

4

Photos:

1, 2 AND **4:** Dried *Pteronisis*. Photos: PA

3: Close-up of axial nodes and internodes. View: ca 4 mm. *Photo: PA*

Zignisis

Alderslade, 1998

Colony shape: Small, profusely branched colonies, growing in one plane or slightly bushy. The main branches grow in a zigzag manner, alternately giving off side branches at each bend. These thinner secondary branches may remain more or less un-branched, but generally they re-branch in a similar zigzag manner. Some species are delicately branched, while others are quite robust. The calcareous internodes are generally 4-sided near the tips of the narrow twigs, but have more than 4 sides nearer to the twigs' base. In the thicker main branches and stem, the internodes have numerous longitudinal ridges. Internodes may or may not have small denticles.

Polyps: Small, monomorphic, and not retractile. They are angled upwards so the mouth faces along or towards the branch surface. Generally they are arranged all around the branches, but they can be biserial on the thinner twigs.

Sclerites: Both the head and body of the polyp are covered in oval or crescent-shaped scales. The outer face of the scales is smooth, and the upper edge may be scalloped, may have a few tooth-like projections, and commonly has a central notch. The basal edge of some of the scales has long root-like structures. The scales of the polyp head and body are separated by a naked neck zone, but in preserved specimens this is usually closed and difficult to see. When the polyp contracts and the tentacles fold in over the mouth, the summit becomes covered by an 8-segmented operculum-like structure. Each triangular segment generally consists of 2 transverse scales followed by a complex arrangement of tubercular rods, clubs, and branched forms. In one species each segment contains a single row of transverse scales of decreasing size. The tentacles contain numerous, small, granular scales that are often curved. The coenenchyme on the surface of the branches contains 2 layers of sclerites. The upper layer contains bulbous, cushion-like sclerites with a mostly smooth undulating upper surface and root-like tubercles on the base. Sometimes the cushion is laterally compressed forming a thick scale with prominent ribs. The lower layer contains ovals and spindles, with complex tubercles often in girdles, tuberculate plates, and capstans. Sclerites are usually rust coloured and look yellowish under a microscope, but sometimes they are colourless.

Colour: Most colonies are a shade of brown. Some have white to yellowish polyps, and a few are totally white. The axial internodes can be shades of yellow, orange and brown, or grey-white. Azooxanthellate.

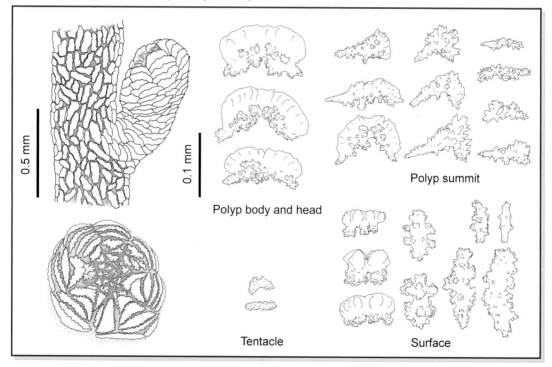

0.5 mm

0.1 mm

Polyp body and head

Polyp summit

Tentacle

Surface

Habitat and abundance:
Very little data is available, as most specimens have been collected by dredge from 12 - 120 m. Those collected by diving have been on rock walls, cave floors, or under ledges.

Zoogeographic distribution:
Only known from Australia, off the southern and western coast as far north as Port Hedland.

Similar Indo-Pacific genera:
Colonies of *Stephanogorgia* have the same growth form, but they do not have a segmented axis.

Photo:
1: Western Australia. *Photo: Graham Edgar*

2: Preserved specimen, 11.5 cm across. *Photo: PA*

3: Close-up of segmented axis, 2 mm long. *Photo: PA*

1

2

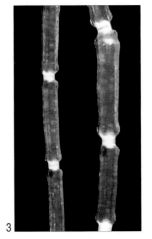

3

LITERATURE

Alderslade, P. 1983. *Dampia pocilloporaeformis*, a new genus and a new species of Octocorallia (Coelenterata) from Australia. *The Beagle* **1**: 33-40

Alderslade, P. 1986. An unusual leaf-like gorgonian (Coelenterata: Octocorallia) from the Great Barrier Reef. *The Beagle* **3**: 81-93

Alderslade, P. 1986. New species of the gorgonian genus *Plumigorgia* (Coelenterata: Octocorallia) with a review of the family Ifalukellidae. *The Beagle, Records of the Northern Territory Museum of Arts and Sciences* **3**: 95-122

Alderslade, P. 1998. Revisionary systematics in the gorgonian family Isididae, with descriptions of numerous new taxa (Coelenterata: Octocorallia). *Records of the Western Australian Museum*, Supplement No. **55**: 359pp.

Alderslade, P. 2000. Four new genera of soft corals (Coelenterata:Octocorallia), with notes on the classification of some established taxa. *Zoologische Mededlingen, Leiden* **74**: 237-249

Allen, G. R., and Steene, R. 1994. *Indo-Pacific coral reef field guide.* Tropical Reef Research, Singapore, 378 pp

Bayer, F. M. 1974. Studies on the anatomy and histology of *Plexaura homomalla* in Florida. In: Bayer, F. M., and A. J. Weinheimer (eds.), *Studies in Tropical Oceanography* **12**: 62-100

Bayer, F. M. 1981. Key to the genera of Octocorallia exclusive of Pennatulacea (Coelenterata: Anthozoa), with diagnoses of new taxa. *Proceedings of the Biological Society of Washington* **94**: 901-947

Bayer, F. M. 1992. The helioporacean octocoral *Epiphaxum*: Recent and fossil. A monographic iconography. *Studies in Tropical Oceanography* **15**: 1-76

Bayer, F. M., Grasshoff, M., and Verseveldt, J. 1983. *Illustrated trilingual glossary of morphological and anatomical terms applied to Octocorallia.* Brill, Leiden, 75 pp

Benayahu, Y. 1991. Reproduction and developmental pathways of Red Sea Xeniidae (Octocorallia: Alcyonacea). *Hydrobiologia* **216**: 125-130

Benayahu, Y., Achituv, Y., and Berner, T. 1988. Embryogenesis and acquisition of algal symbionts by planulae of *Xenia umbellata* (Octocorallia: Alcyonacea). *Marine Biology* **100**: 93-101

Benayahu, Y., Achituv, Y., and Berner, T. 1989. Metamorphosis of an octocoral primary polyp and its infection by algal symbiosis. *Symbiosis* **7**: 159-169

Benayahu, Y., and Loya, Y. 1981. Competition for space among coral reef sessile organisms at Eilat, Red Sea. *Bulletin of Marine Science* **31**: 514-22

Benayahu, Y., and Loya, Y. 1984. Life history studies on the Red Sea soft coral *Xenia macrospiculata* Gohar, 1940. I. Annual dynamics of gonadal development. *Biological Bulletin* **166**: 32-43

Benayahu, Y., Weil, D., and Malik, Z. 1992. Entry of algal symbionts into oocytes of the coral *Litophyton arboreum. Tissue and Cell* **24**: 473-482

Brazeau, D. A., and Lasker, H. R. 1990. Sexual reproduction and external brooding by the Caribbean gorgonian *Briareum asbestinum. Marine Biology* **104**: 465-474

Coll, J. C. 1992. The chemistry and chemical ecology of octocorals (Coelenterata, Anthozoa, Octocorallia). *Chemical Review* **92**: 613-631

Coll, J. C., Bowden, B. F., Heaton, A., Scheuer, P. J., Li, M., Clardey, J., Schulte, G. K., and Finer-Moore, J. 1989. Structures and possible functions of epoxypukalide and pukalide diterpenes associated with eggs of sinularian soft corals (Cnidaria, Anthozoa, Octocorallia, Alcyonacea, Alcyoniidae). *Journal of Chemical Ecology* **15**: 1177-1191

Coll, J. C., La Barre, S., Sammarco, P. W., Williams, W. T., and Bakus, G. J. 1982b. Chemical defences in soft corals (Coelenterata: Octocorallia) of the Great Barrier Reef: A study of comparative toxicities. *Marine Ecology Progress Series* **8**: 271-278

Coll, J. C., Leone, P. A., Bowden, B. F., Carrol, A. R., König, G. M., Heaton, A., Nys, R. D., Maida, M., Alino, P. M., Willis, R. H., Babcock, R. C., Florian, Z., Clayton, M. N., Miller, R. L., and Alderslade, P. N. 1995. Chemical aspects of mass spawining in corals II. - Epi-thunbergol, the sperm attractant in the eggs of the soft coral *Lobophytum crassum. Marine Biology* **123**: 137-143

Coma, R., Gili, J. M., Zabala, M., and Riera, T. 1994. Feeding and prey capture cycles in the aposymbiontic gorgonian *Paramuricea clavata*. *Marine Ecology* **115**: 257-270

Dahan, M., and Benayahu, Y. 1997. Clonal propagation by the zooxanthellate octocoral *Dendronephthya hemprichi*. *Coral Reefs* **16**: 5-12

Dai, C. F. 1990. Interspecific competition between Taiwanese corals with special reference to interactions between alcyonaceans and scleractinians. *Marine Ecology Progress Series* **60**: 291-297

De'ath, G., and Moran, P. 1998. Factors affecting behaviours of crown-of-thorns starfish (*Acanthaster planci*). 2: Feeding preferences. *Journal of Experimental Marine Biology and Ecology* **220**: 107-126

Devantier, L.M., De'ath, G., Done, T. J., and Turak, E. 1998. Ecological assessment of a complex natural system: a case study from the Great Barrier Reef. *Ecological Applications* **8**: 480-496

Dinesen, Z. D. 1983. Patterns in the distribution of soft corals across the central Great Barrier Reef. *Coral Reefs* **1**: 229-236

Done, T. J. 1982. Patterns in the distribution of coral communities across the central Great Barrier Reef. *Coral Reefs* **1**: 95-107

Done, T. J. 1992. Phase shifts in coral communities and their ecological significance. *Hydrobiologia* **247**: 121-132

Done, T. J. 1999. Coral community adaptability to environmental change at the scales of regions, reefs and reef zones. *American Zoologist* **39**: 66-79

Done, T. J., Dayton, P. K., Dayton, A. E., and Steger, R. 1991. Regional and local variability in recovery of shallow coral communities: Moorea, French Polynesia and central Great Barrier Reef. *Coral Reefs* **9**: 183-192

English, S., Wilkinson, C., and Baker, V. 1999. *Survey manual for tropical marine resources*. 2nd edition. Australian Institute of Marine Science, Townsville, 402 pp

Fabricius, K. E. 1997. Soft coral abundance in the central Great Barrier Reef: effects of *Acanthaster planci*, space availability and aspects of the physical environment. *Coral Reefs* **16**: 159-167

Fabricius, K. E. 1998. Reef invasion by soft corals: Which taxa and which habitats? Pp 77-90 *in:* Greenwood, J. G., and Hall, N. J. (eds.) *Proceedings of the Australian Coral Reef Society 75th Anniversary Conference, Heron Island, October 1997:* School of Marine Science, The University of Queensland, Brisbane

Fabricius, K. E. 1999. Tissue loss and mortality in soft corals following mass-bleaching. *Coral Reefs* **18**: 54

Fabricius, K. E. and De'ath, G. 1997. The effects of flow, depth and slope on cover of soft coral taxa and growth forms on Davies Reef, Great Barrier Reef. Pp 1071-1076 *in:* Lessios, H. A. (ed.) *Proceedings of the 8th International Coral Reef Symposium, Panama,* Vol. 2. Smithsonian Tropical Research Institute, Balboa, Republic of Panama

Fabricius, K. E., and De'ath, G. 2000 *Soft Coral Atlas of the Great Barrier Reef.* Australian Institute of Marine Science, http://www.aims.gov.au/softcoral.atlas : 57 pp

Fabricius, K. E., and De'ath, G. 2001. Biodiversity on the Great Barrier Reef: Large-scale patterns and turbidity-related local loss of soft coral taxa. Pp 127 - 144 *in:* Wolanski, E. (ed.) *Oceanographic processes of coral reefs: physical and biological links in the Great Barrier Reef.* CRC Press, London, 356 pp

Fabricius, K. E., and Dommisse, M. 2000. Depletion of suspended particulate matter over coastal reef communities dominated by zooxanthellate soft corals. *Marine Ecology Progress Series* **196**: 157-167

Fabricius, K. E., Benayahu, Y., and Genin, A. 1995a. Herbivory in asymbiotic soft corals. *Science* **268**: 90-92

Fabricius, K. E., Genin, A., and Benayahu, Y. 1995b. Flow-dependent herbivory and growth in zooxanthellae-free soft corals. *Limnology and Oceanography* **40**: 1290-1301

Fabricius K. E., and Klumpp, D. W. 1995. Wide-spread mixotrophy in reef-inhabiting soft corals: the influence of depth, and colony expansion and contraction on photosynthesis. *Marine Ecology Progress Series* **125**: 195-204

Fosså, S. A., and Nilsen, A. J. 1998. The Modern Coral Reef Aquarium, Volume 2. Birgit Schmettkamp Verlag, Bornheim, Germany

Garzón-Ferreira, J., and Zea, S. 1992. A mass mortality of *Gorgonia ventalina* (Cnidaria: Gorgoniidae) in the Santa Marta area, Caribbean coast of Colombia. *Bulletin of Marine Science* **50**: 522-526

Gohar, H. A. F. 1939. On a new xeniid genus *Efflatounaria. Annual Magzine of Natural History* **(11) 3**: 32-36

Gosliner, T. M., Behrens D. W., and Williams, G. C. 1996. *Coral reef animals of the Indo-Pacific.* Sea Challengers, Monterey, California, 314 pp

Grasshoff, M. and Alderslade, P. 1997. Gorgoniidae of Indo-Pacific reefs with descriptions of two new genera (Coelenterata: Octocorallia). *Senckenbergiana Biologica* **77**: 23-25.

Grasshoff, M. 1999. The shallow water gorgonians of New Caledonia and adjacent islands (Coelenterata, Octocorallia). *Senckenbergiana Biologica* **78**: 121 pp

Grasshoff, M. 2000. The gorgonians of the Sinai coast and the Strait of Gubal, Red Sea (Coelenterata, Octocorallia). *Courier Forschungsinstitut Senckenberg* **224**: 1-12

Hoegh-Guldberg O. 1999. Climate change, coral bleaching, and the future of the world's coral reefs. *Marine and Freshwater Research* **50**: 839-866

Hughes, T. P., and Connell, J. H. 1999. Multiple stressors on coral reefs: A long-term perspective. *Limnology and Oceanography* **44**: 932-940

Hyman, L. H. 1940. *The Invertebrates. Protozoa through Ctenophora.* McGraw-Hill, New York, 716 pp

Johnson, D. P., and M. J. Risk. 1987. Fringing reef growth on a terrigenous mud foundation, Fantome Island, central Great Barrier Reef, Australia. *Sedimentology* **34**: 275-287

Jones, R., Hoegh-Guldberg, O., Larkum, A. W. L., and Schreiber, U. 1998. Temperature induced bleaching of corals begins with impairment of dark metabolism in zooxanthellae. *Plant Cell and Environment* **21**: 1219-1230

Kleypas, J. A. 1996. Coral reef development under naturally turbid conditions: fringing reefs near Broad Sound, Australia. *Coral Reefs* **15**: 153-167

Konishi, K. 1981. Alcyonarian spiculite: limestone of soft corals. *Proceedings of the 4th International Coral Reef Symposium,* Vol. **1:** 643-649

Lasker, H. R., and Kim, K. 1996. Larval development and settlement behavior of the gorgonian coral *Plexaura kuna* (Lasker, Kim and Coffroth). *Journal of Experimental Marine Biology and Ecology* **207**: 161-175

Lewis, J. B. 1982. Feeding behaviour and feeding ecology of the Octocorallia (Anthozoa). *Journal of Zoology, London* **196**: 371-384

Maida, M., Sammarco, P. W., and Coll, J. C. 1995. Preliminary evidence for directional allelopathic effects of the soft coral *Sinularia flexibilis* (Alcyonacea: Octocorallia) on scleractinian coral recruitment. *Bulletin of Marine Science* **56**: 303-311

Mariscal, R. N., and Bigger, C. H. 1977. Possible ecological significance of octocoral epithelial ultrastructure. *Proceedings of the 3rd International Coral Reef Symosium, Miami*: 127-133

Marsh, L. M., R. H. Bradbury, and R. E. Reichelt. 1984. Determination of the physical parameters of coral distribution using line transect data. *Coral Reefs* **2**: 175-180

Michalek-Wagner, K. and Willis, B. L. 2001. Impacts of bleaching on the soft coral *Lobophytum compactum.* I. Fecundity, fertilisation and offspring viability. *Coral Reefs* **19**: 231-239

Nagelkerken, I., Buchan, K., Smith, G. W., Bonair, K., Bush, P., Garzón-Ferreira, J., Botero, L.,Gayle, P., Heberer, C., Petrovic, C., Pors, L., and Yoshioka, P. 1997. Widespread disease in Caribbean sea fans: I. Spreading and general characteristics. *Proceedings of the 8th International Coral Reef Symposium* **1**: 679-682

Peters, Esther C. 1992. The role of environmental stress in the development of coral diseases and micro-parasite infestations. *American Zoologist* **32**: 960.

Ofwegen van, L.P. 1987. Melithaeidae (Coelenterata: Anthozoa) from the Indian Ocean and the Malay Archipelago. *Zoologische Verhandelingen* **239**: 1-57

Ofwegen van, L.P., Goh, N.K.C., and Chou, L.M. 2000. The Melithaeidae (Coelenterata: Octocorallia) of Singapore. *Zoologisch Meded. Leiden* **73** (19): 285-304

Ofwegen van, L.P. and Schleyer, M. H. 1997. Corals of the South-west Indian Ocean V. *Leptophyton benayahui* gen. nov. & spec. nov. (Cnidaria, Alcyonacea) from deep reefs off Durban and on the KwaZulu-Natal south coast, South Africa. *South African Association for Marine Biological Research, Oceanographic Research*

Institute Investigational Report **71**: 1-12

Pass, M. A., Capra, M. F., Carlisle, C. H., Lawn, I., and Coll, J. C. 1989. Stimulation of contractions in the polyps of the soft coral *Xenia elongata* by compounds extracted from other alcyonacean soft corals. *Comparative Biochemistry and Physiology* **94C**: 677-81

Ribes, M., R. Coma, and J. M. Gili. 1998. Heterotrophic feeding by gorgonian corals with symbiotic zooxanthella. *Limnology and Oceanography* **43**: 1170-1179

Riegel, B., and Branch, G. M. 1995. Effects of sediment on the energy budgets of four scleractinian (Bourne 1900) and five alcyonacean (Lamouroux 1816) corals. *Journal of Experimental Marine Biology and Ecology* **186**: 259-275

Sammarco, P. W., and Coll, J. C. 1997. Secondary metabolites or primary? Re-examination of a concept through a marine example. Pp 1245-1250 *in:* Lessios, H. A. (ed.) *Proceedings of the 8th International Coral Reef Symposium, Panama,* Vol. 2. Smithsonian Tropical Research Institute, Balboa, Republic of Panama

Sammarco, P. W., Coll, J. C., La Barre, S., and Willis, B. 1983. Competitive strategies of soft corals (Coelenterata: Octocorallia): Allelopathic effects on selected scleractinian corals. *Coral Reefs* **1**: 173-178

Schlichter, D. 1982a. Nutritional strategies of cnidarians: the absorption, translocation and utilization of dissolved nutrients by *Heteroxenia fuscescens*. *American Zoologist* **22**: 659-69

Schlichter, D. 1982b. Epidermal nutrition of the alcyonarian *Heteroxenia fuscescens* (Ehrb.): Absorption of dissolved organic material and lost endogenous photosynthates. *Oeceologia* **53**: 40-9

Schmidt, H., and Moraw, B. 1982. Die Cnidogenese der Octocorallia (Anthozoa, Cnidaria): II Reifung, Wanderung und Zerfall von Cnidoblast und Nesselkapsel. *Helgoländer Meeresuntersuchungen* **35**: 97-118

Schuhmacher, H. 1997. Soft corals as reef builders. pp 499-502 *in:* Lessios, H. A. (ed.) *Proceedings of the 8th International Coral Reef Symposium, Panama,* Vol. 2. Smithsonian Tropical Research Institute, Balboa, Republic of Panama.

Sebens, K. P., and Mills, J. S. 1988. Sweeper tentacles in a gorgonian octocoral: morphological modifications for interference competition. *Biological Bulletin* **175**: 378-387

Slattery, M., Hines, G. A., Starmer, J., and Paul, V. J. 1999. Chemical signals in gametogenesis, spawning, and larval settlement and defense of the soft coral *Sinularia polydactyla*. *Coral Reefs* **18**: 75-84

Smith, G., Ives, L. D., Nagelkerken, I. A., and Ritchie, K. B. 1996. Caribbean sea fan mortalities. *Nature* **383**: 487

Sprung, J., and Delbeek, J. C. 1997. *The reef aquarium*. Volume 2. Ricordea Publishing, Inc., Coconut Grove, Florida, USA, 546 pp

Tixier-Durivault, A. 1987. Sous-classe des octocorallieres (Octocorallia Haeckel, 1866; Octactinia Ehrenberg, 1828; ... In: Doumence, D. (ed.), Traité de Zoologie, Tome 3, Fasc. 3, Cnidaires, Anthozoaires. Paris, France

Tursch, B., and Tursch, A. 1982. The soft coral community on a sheltered reef quadrat at Laing Island (Papua New Guinea). *Marine Biology* **68**: 321-332

Veron, J. E. N. 1995. *Corals in space and time*. University of New South Wales Press, Sydney, 321 pp

Veron, J. E. N. 2000. *Corals of the World*. Volume 1 – 3. Staffort-Smith, M. (ed.) Australian Institute of Marine Science, Townsville, Australia.

Veron, J. E. N., and Pichon, M. 1982. *Scleractinia of Eastern Australia. Part 4, Family Poritidae*. Australian Institute of Marine Science Monograph Series **V**, 210 pp

Verseveldt, J. 1980. A revision of the genus *Sinularia* May (Octocorallia, Alcyonacea). *Zoologische Verhandelingen, Leiden* **179**: 1-128

Verseveldt, J. 1982. A revision of the genus *Sarcophyton* Lesson (Octocorallia, Alcyonacea). *Zoologische Verhandelingen, Leiden* **192**: 1-91

Verseveldt, J. 1983 A revision of the genus *Lobophytum* von Marenzeller (Octocorallia, Alcyonacea). *Zoologische Verhandelingen, Leiden* **200**: 1-103

Verseveldt, J. 1983. The octocorallian genera *Spongodes* Lesson, *Neospongodes* Kükenthal and *Stereonephthya* Kükenthal. *Beaufortia* **33**: 1-13

Verseveldt, J., and Bayer, F. M. 1988. A revision of the genera *Bellonella, Eleutherobia, Nidalia* and *Nidaliopsis* (Octocorallia: Alcyoniidae and Nidaliidae), with descriptions of two new genera. *Zoologische Verhandelingen, Leiden* **245**: 1-131

Wallace, C. C. 1999. *Staghorn corals of the world.* CSIRO Publ., Melbourne, 422 pp

Wilkinson, C. R. 2000. *Status of coral reefs of the world: 2000.* Australian Institute of Marine Science, Townsville, 364 pp

Williams, G. C. 1992a. Biotic diversity, biogeography, and phylogeny of pennatulacean octocorals associated with coral reefs in the Indo-Pacific. *Proceedings of the 7th International Coral Reef Symposium, Guam* **2**: 729-735

Williams, G. C. 1992b. Revision of the soft coral genus *Minabea* (Octocorallia: Alcyoniidae) with new taxa from the Indo-West Pacific. *Proceedings of the Californian Academy of Sciences* **48**: 1-26

Williams, G. C. 1993. Coral Reef Octocorals. *An illustrated guide to the soft corals, sea fans and sea pens inhabiting the coral reefs of northern Natal.* Natural Science Museum, Durban, 64 pp

Williams, G. C. 1995. Living genera of sea pens (Coelenterata: Pennatulacea): illustrated key and synopses. *Zoological Journal of the Linnean Society* **113**: 93-140

Williams, G. C. 1999. Index Pennatulacea. Annotated bibliography and indexes of the sea pens (Coelenterata: Octocorallia) of the World 1469–1999. *Proceedings of the Californian Academy of Sciences* **51**: 19-103

Williams, G. C., and Alderslade, P. 1999. Revisionary systematics of the western Pacific soft coral genus *Minabea* (Octocorallia: Alcyoniidae), with descriptions of a related new genus and species from the Indo-Pacific. *Proceedings of the Californian Academy of Sciences* **51**: 337-364

Zaslow, R. B., and Benayahu, Y. 1996. Longevity, competence and energetic content in planulae of the soft coral *Heteroxenia fuscescens. Journal of Experimental Marine Biology and Ecology* **206**: 55-68

Isis hippuris on a back reef crest of the Great Barrier Reef. *Photo: KF*

Katharina Fabricius is a marine biologist who has studied the ecology of coral reefs since 1988. Within a couple of years she became intrigued about how little was known about octocorals, despite their taxonomic richness and high abundances. Since then, she has has published numerous research papers and has written book chapters on the biology and ecology of octocorals and other reef-inhabiting organisms, and on the effects of the physical environment and human activities on the biodiversity of coral reefs. Her PhD was awarded by the University of Munich for research on nutrition and community regulation in tropical soft corals in 1995. In the meantime, she has conducted over 1500 taxonomic inventories of octocorals as part of community surveys on the Great Barrier Reef, the Indian and Pacific Ocean, the Red Sea, and the Florida Keys. She holds a position as Research Scientist at the Australian Institute of Marine Science. Having grown up in Germany, she is now hooked on living on a small island off Townsville in tropical Australia.

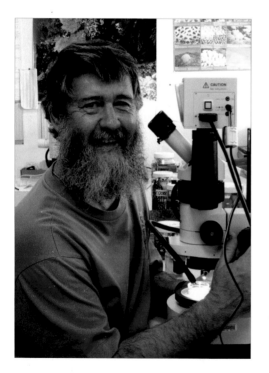

Phil Alderslade began working on the identification of octocorals nearly 30 years ago, when he was a graduate research assistant in the Zoology Department of the University of Queensland. He continued to specialise in this field while working for the Roche Research Institute of Marine Pharmacology in Sydney for most of the 1970's, and also as the Curator of Coelenterates at the Museum and Art Gallery of the Northern Territory, Darwin: a post he has held since 1981. Phil's research gained him an MSc and a PhD from James Cook University, and he is now considered a world authority in octocoral taxonomy, having contributed chapters to a number of books and published many scientific papers. He regularly works as taxonomic consultant to international research organisations, and is on the editorial committee of a scientific journal. He is married with two grown up children, likes fishing, dabbles in opal cutting and stained glass, and confesses at times to eating too much chocolate.